Britain's wild harvest

The commercial uses of wild plants and fungi

By Hew D.V. Prendergast and Helen Sanderson
Photographs by Andrew McRobb

This book is dedicated to

EDVP
Whose elderflower wine,
freely given,
was peerless

Ange and Irv
For your enthusiasm, inspiration and love for nature

First published 2004

Printed in China by Compass Press Ltd
for the Publishers
Royal Botanic Gardens, Kew
Richmond, Surrey TW8 3AE

ISBN 1 84246 072 2

contents

acknowledgements

This book derives from a report, *Commercial uses of wild and traditionally managed plants in England and Scotland*, submitted in early 2002 to its sponsors, the Countryside Agency, English Nature and Scottish Natural Heritage. Subsequently we extended our coverage to Wales, thanks to support from the Atlantic Foundation and the Thriplow Charitable Trust. We are indebted to these organisations, and especially to David Gear, Rob Green and Richard Lloyd of the Countryside Agency. Countless people contributed to those reports and to this book. Some are mentioned here, but many are not; to them all we proffer our thanks. A special recognition goes to the harvesters, producers and traders for their great patience, co-operation and enthusiasm.

Here at Kew we are particularly grateful to: Andrew McRobb, for his dedication and for taking most of the photographs; Jeff Eden, for the design; Peter Roberts for his mycological expertise; and our colleagues in the Centre for Economic Botany who put up with ever growing piles of British products in our offices and then, eventually, shared with us the pleasure in eating, drinking or simply touching them.

foreword

More than ever before, the British countryside is in the news, and it is not hard to see why. Livestock disease, agricultural policy, planning concerns, the loss of farmland birds, hunting and many other issues have all drawn our attention to a fundamental question: what exactly do we want from our countryside?

The countryside that we have today bears the heavy print of history. Forests have been cleared, fens drained and other habitats like hedgerows and meadows, once characteristic of much of Britain, have been created. How we want our countryside to look like in future is in our hands. If it is to be productive and profitable, and yet remain beautiful and rich in wildlife, we must support those who work there in many ways – including through buying their products.

In this book we learn about a very specific range of products – those based on our wild plants, fungi and algae. We also learn about people – the associated collectors, producers and traders who are passionate defenders of the places where they live and work. This book is a product of almost two years research carried out at the Royal Botanic Gardens, Kew, supported by the Countryside Agency, both government agencies funded by Defra. The work took the authors the length and breadth of Britain, from Devonshire moors and Norfolk salt marshes, to Cumbrian woodlands, rocky shores of Mull and slow-moving rivers of Bedfordshire. Here they interviewed, observed and recorded the diversity of our often rather overlooked contemporary rural economy.

Managing our environment in a sustainable way is the mutual goal of both the Countryside Agency and the Royal Botanic Gardens, Kew. We hope that this book – the first for over half a century to look at the economic role of our wild flora – will stimulate a greater interest in the wonderful goods that our countryside can produce and show how this can be done in sustainable ways without compromising the future of our equally wonderful wildlife.

Peter R. Crane FRS, Director, Royal Botanic Gardens, Kew
Sir Ewen Cameron, Chairman, The Countryside Agency

introduction

A cone of woven willow branches stands against the wall of our open plan office. About four feet high, it looks like one of those pea cones sold by garden centres. But what actually is it, and where does it come from? Visitors often try to guess. With some prompting they come to recognise it as a fish-trap – for its shape is almost universal – but as to its origin there is no such consensus: New Guinea, perhaps, or up the Amazon or in Africa, almost always somewhere remote and tropical. Then the more perceptive notice strands of seaweed tangled among the ribs of the cone and it is time to reveal the answer: the Severn estuary.

1

Salmon putcher from the Severn estuary.

For centuries local fishermen have set long lines of these so-called 'putchers' to catch salmon migrating to their spawning grounds. Due to overfishing in the Atlantic, pollution and other causes, the salmon has declined catastrophically and by 1995 just a handful of licenses were issued for trapping. While the rest of the Severn fishermen were by then using aluminium putchers, one continued to make them from the traditional material of almond willow (*Salix triandra*). He kindly donated the one now standing in our office, and showed us the very trees, on a lane side verge, from which he cut the branches to make it.

As years pass, this putcher will remain as one of the few tangible vestiges of an extinct way of life. Around the Severn as elsewhere, farmed salmon have supplanted wild ones in our diet. Only for sport have the wild fish retained economic importance, on the River Tweed alone supporting some 500 jobs. Red deer and red grouse, Britain's other prize game, have also succumbed to increasing manipulation. The red deer, the so-called Monarch of the Glen, has become a farm animal while the heather-dominated moors on which grouse depend are intensely managed.

Such succumbing to our intervention has also of course been the fate of thousands of plant and animal species all over the world. In Britain we have reaped the benefits. The wetter west and north concentrates on the raising of cattle and sheep originally domesticated thousands of years ago in Europe and the Middle East. By contrast the drier climate and higher-grade soils of East Anglia and Lincolnshire, for example, grow cereals from the same region, as well as potatoes (*Solanum tuberosum*) from the Andes, maize (*Zea mays*) from Central America, and peas (*Pisum sativum*) from the Mediterranean. Elsewhere experiments are attempting to grow two unlikely plants from China: tea (*Camellia sinensis*) and soya (*Glycine max*). The orchards of Devon, Kent, Somerset and the Vale of Evesham contain descendants of apples (*Malus domestica*) that arrived via the fabled Silk Route from Central Asia, and in our newest plantations are Sitka spruce (*Picea sitchensis*) and Douglas fir (*Pseudotsuga menziesii*) from the Pacific coast of North America, now among the tallest trees in our landscape. On these imports, and others, Britain has depended and thrived.

But the slow and dispersed revolution that saw all these plants taken out of the wild and into cultivation did not bypass species native to Britain. On the cliffs of southern England and Wales sprouts the wild ancestor of cabbage, cauliflower and broccoli (*Brassica oleracea*). Sea kale (*Crambe maritima*), an attractive plant with an erstwhile culinary reputation, decorates some of Britain's bare and inhospitable shingle beaches. Unspectacular it may be but the sea beet (*Beta vulgaris*) has given rise to many varieties, including the one grown mainly in East Anglia for its extraordinarily high levels of sugar. Other coastal species like carrot (*Daucus carota*) and parsnip (*Pastinaca sativa*) have become important vegetables, although their cultivation started elsewhere, while thrift (*Armeria maritima*) has entered our gardens for the same pink flowers that swathe our cliff tops.

Native shrubs have also been developed. Mezereon (*Daphne mezereum*), a plant of calcareous woodlands, is widely grown for its early spring scent while domesticated descendants of gooseberry (*Ribes uva-crispa*), red currant (*R. rubrum*) and black currant

Sea kale, one of our most striking coastal plants, growing on the shingle beach at Dungeness (Kent).
PHOTOGRAPH: HEW PRENDERGAST

(*R. nigrum*) are the basis of a lucrative soft fruit industry. The raspberry (*Rubus idaeus*), another woodland plant, has found a commercial epicentre around one city, Dundee, whose fruit-processing expertise started with the transformation of bitter Spanish oranges (*Citrus aurantium*) into what is now a well known marmalade.

Some of our native species have attained economic prominence elsewhere because of their curative properties. Due to its success in the treatment of depression, perforate St John's Wort (*Hypericum perforatum*) is grown on a huge scale in the USA, and one of our most weedy plants, the greater plantain (*Plantago major*), has established itself as a staple of herbal medicine in the New World from Brazil to New York and beyond. Elder (*Sambucus nigra*) is still expanding its range among tribal peoples of Amazonia for its use in a wide range of complaints.

Other native species have provided food, medicine, and horticultural splendour. For them management alone was enough, needing neither cultivation nor selective breeding over generations. In this sense, our main hardwood tree species are still largely wild despite their importance as timber and fuel, and their dominance of the British landscape. Although hazel (*Corylus avellana*) has followed the domestication path of beet and cabbage because of its edible nuts, it has not done so for its more critical role in the millennia-old cycle of coppicing (regular cutting to produce straight poles). Some species have escaped attention altogether – brown seaweeds, for example, despite their once vital role as fertiliser on many of our islands from Scilly to Shetland, and fungi because, it seems, they were simply too fearful to contemplate. Now our views, use and appreciation of such species have changed.

In the last 40 years Britain has been a world leader in understanding the distribution of its wildlife. Botanists pioneered the way by mapping our flora by 1962 (and again in 2002), and since then everything from dragonflies to dormice has been searched for, counted, and recorded, often in so-called Biodiversity Action Plans compiled for their survival. But far less is known of the extent to which wild species still support people's livings or provide some form of income. In 1951 a well-known forester, Herbert Edlin[1], wrote a book on *British plants and their uses*. He described a host of what were then commonplace activities: for example the use of heather (*Calluna vulgaris*) in the foundations of roads and runways, the large scale collecting of brown seaweeds from the Scottish west coast for alginates, the cultivation in Lincolnshire of fat hen (*Chenopodium bonus-henricus*) for its tender shoots, and the gathering of broom (*Cytisus scoparius*) for its medicinally important ingredients.

While these practices have gone, others have continued due to deep roots in local culture or because they have retained, or regained, an economic rationale. At the same time new fashions and interests have arisen. Celebrity chefs stimulate the collecting of wild plants and fungi while supermarket shelves tempt consumers with ever more foods from all over the world. Some consumers are increasingly interested in knowing about the chain of supply – about where the food comes from, how it is produced, and the distance it has travelled. There is a growing

[1] Edlin, H.L. 1951. *British plants and their uses*. London: B.T. Batsford.

market for forest or woodland products bearing labels of ethical and sustainable methods of production, and for goods that are definitive and typical – even part of the heritage – of a certain county or region. In some such areas, rural livelihoods are also a concern given that modern agriculture needs less and less jobs. Echoing similar initiatives elsewhere in Europe, the Countryside Agency and other organisations have been active in promoting countryside products – "products which originate from environmentally responsible land management and which can encourage local economic activity, or a sense of local or regional identity"[2].

In this book we look at the current role of wild plants, fungi and algae in the economy. Like Edlin, we exclude timber species like beech (*Fagus sylvatica*), grown in the Chilterns for furniture, Scots pine (*Pinus sylvestris*), planted in the uplands for planking, and the anciently introduced sweet chestnut (*Castanea sativa*) that provides Kent and Sussex with its characteristic fencing. We are also flexible with our use of the term 'wild'. For example, we include those willows that, despite being developed into a number of varieties, have become an essential part of the Somerset Levels and, through basketry, given rise to distinctive local skills and quintessential countryside products. Willows planted for biofuel (so-called Short Term Coppice) do not appear, however, and neither do all those species being investigated or developed as future sources of industrial materials.

This book therefore takes us to the corners and countryside of Britain where we find wild species in commercial use: the mudflats of Norfolk, the rocky shores of Argyll and Devon, the moors of Shropshire, and even into Swansea market. It features the wild (or wilder) species whose hold on us has neither been usurped by domesticated descendants, synthetic replacements nor by rivals from elsewhere and meets the people who exploit them and fashion them into products for the market. To adapt the name of its cultural equivalent written by Richard Mabey[3], it is a British Flora Economica, and a snapshot of a continuing evolving relationship, based on utility, between the flora and peoples of Britain.

[2] For more information on the Countryside Agency's work on countryside products, visit its website at www.countryside.gov.uk. and look under 'eat the view'.

[3] Mabey, R. 1996. *Flora Britannica*. London: Sinclair-Stevenson.

Pollarded willows on the Somerset Levels.

food and drink

For many people, collecting and consuming wild plants has been part and parcel of country life since their childhood. For others, the allure is more recent, and commercial enterprises the length and breadth of Britain have sprung up to cater for the demand. From seashore to uplands, Britain has a host of edible wild plants, ready to be transformed into wrappings, sauces, condiments, jams, pies, cordials, beers, wines, and liqueurs. Some of the more popular species have even been taken into cultivation.

2

Cornish Yarg cheese in the process of being 'nettled'.

food

Throughout Britain, some 2,000 people are involved in the collecting, processing and selling of wild species for food and drinks. Several companies have a 'broad brush' approach, not specialising in any particular plant or habitat. How much and where they collect, and the values of their sales are of course sensitive information, but one of the biggest companies working in this way is Caledonian Wildfoods Ltd in Bellshill (Strathclyde). With 200 seasonal pickers, it gathers the leaves of a wide range of plants like watercress (*Rorippa nasturtium-aquaticum*), bog myrtle (*Myrica gale*) and sorrel (*Rumex acetosa*) and the fruits of others like crab apple (*Malus sylvestris*), hawthorn (*Crataegus monogyna*) and rowan (*Sorbus aucuparia*) to supply shops and restaurants in Britain and abroad. In spring and early summer it exports about 500 kilos a week of the leaves of wild garlic or ramsoms (*Allium ursinum*) to the USA.

Left: **Wild garlic leaves can be used in salads and soups, and are particularly good with snails, sea bass, or wrapped around lamb before roasting. Lynher Dairies (Cornwall), producers of Cornish Yarg cheese, have recently started producing a cheese wrapped in wild garlic leaves.**

Below: **Brigitte Tee collecting wild garlic in the New Forest. With the permission of landowners she picks the young green leaves of this pungent plant through the summer. Each week during the season Brigitte sells 10 kilos to local restaurants and at some of London's food markets.**

Fungi

Fungi are one of the mainstays of several enterprises dealing with wild foods. Most fungi grow in the form of microscopic filaments called hyphae that extend and branch to form a network or mycelium. Arising from the mycelium are fruiting bodies in the familiar form of mushrooms and toadstools, or a variety of brackets, cups, horns, cushions and gelatinous blobs. Fungi play a vital role in nutrient re-cycling, and form symbiotic associations with plants without which the plants could not survive. Of the 12,000 species that occur in Britain, only a few have a commercial value because of their edibility.

Picking and eating wild fungi is a pleasure enjoyed by relatively few people. Historically, Britons have regarded them more in fear than as food, and our collecting has been so confined that we have had no need for a network of pharmacists, as in France, specially trained to distinguish between the deadly and the delectable. Television chefs, and a growing appreciation of 'food for free', have persuaded us otherwise and now there is an homegrown collecting trade, based on about 20 species.

The collecting season has a short spurt in spring, and returns during the cooler, moister months of autumn. Different species have different gastronomical qualities. Cep, for example, is considered to be one of the tastiest species (see p. 15 for scientific names), while chanterelle and hedgehog mushroom are particularly good with scrambled eggs and omelette, and flavoursome field blewit spices up beef or venison stews.

Britain's largest enterprise trading in wild fungi is McPherson Atlantic Ltd of Tomintoul (Moray). Their core of over a hundred pickers collects some 50 tonnes a year from woods and plantations, mainly in the Highlands. Top pickers reputedly earn up to £30,000 a season, but for most the income is no more than a useful top-up. In the New Forest, one of England's prime spots, income of £2,000 a week is possible for experienced pickers in a good season. The work is not for everyone: early morning starts, an intimate knowledge of locations and habitats, and first class identification skills. Not everyone is up to the task, and in fact the National Poisons Information Service at Guy's and St Thomas' Hospital Trust in London handles over 100 cases of poisoning by fungi every year – although fortunately deaths are very rare.

The right weather is critical. Dry conditions mean that many species simply fail to produce what the collectors are looking for. For species such as puffball, beefsteak and chicken of the woods, supplies may be even more sporadic since for many collectors they are just incidental sidelines to more covetable cep, chanterelle and other 'first division' fungi.

Wild fungi, including oyster mushroom and chanterelle, make a superior filling for a simple omelette.

Main fungi species collected commercially in Britain

Common name	Scientific name
Boletes, including the cep (penny bun)	*Boletus* species, including *B. edulis*
St George's mushroom	*Calocybe gambosa*
Chanterelle	*Cantharellus cibarius*
Velvet shank	*Flammulina velutipes*
Hedgehog mushroom	*Hydnum repandum*
Deceivers	*Laccaria laccata*
Saffron milk-cap	*Lactarius deliciosus*
Chicken of the woods or sulphur polypore	*Laetiporus sulphureus*
Giant puffball	*Langermannia gigantea*
Field blewit, blue leg	*Lepista saeva*
Fairy ring champignon	*Marasmius oreades*
Morel	*Morchella esculenta*
Oyster mushroom	*Pleurotus ostreatus*

Why are the British afraid of fungi?

The mycophobia of the British, both Celts and Anglo-Saxons, is summed up by the word 'toadstool', applied indiscriminately to some 2,000 different species of larger fungi native to the British Isles. Toads, of course, were traditionally considered venomous and so were toadstools, well-known 'to growe where olde rustie iron lieth, or rotten clouts, or neere to serpents dens'. All of them were 'poysonous damp weeds', according to the old herbalists, and best avoided.

The British, however, are not alone in their mistrust of fungi as food items. Worldwide, there seem to be as many cultures that shun them, as those that value them. The reasons are presumably to do with ancient taboos, of the same sort that prevent us from eating horses. Mycophobic peoples, including the British, typically associate fungi with dung, death, and decomposition, and there is enough truth in this for the taboo to be continually reinforced.

For some reason, field mushrooms (*Agaricus campestris*) have long been exempt from the taboo, though the taste for these may have come from France, together with the word 'mousseron' (the etymology and history are lost). A few brave souls even valued blewits, sold in Covent Garden till the nineteenth century and in the Midlands within living memory. But beyond that, the eating of wild fungi seems to have been confined till recently to a few eccentric gentlemen, and even now is considered a strange and dangerous pastime.

There are certainly poisonous fungi out there, just as there are poisonous plants, so anyone thinking of food for free needs to know exactly what they are doing. The safe option for taboo-breakers is to avoid the serpents' dens and stay in the supermarket.

PETER ROBERTS, MYCOLOGY SECTION, RBG KEW

Chicken of the woods or sulphur polypore is a thick and fleshy edible bracket fungus, frequently found on live or decaying oak (*Quercus* species) trees between late spring and autumn.

Every species commands its own price. Per kilo, collectors receive about £3.50 for chanterelle, £4.00 for chicken of the woods and £6 for cep, but by the time the fungi reach the London market the figures escalate: cep, for example, selling for £100 fresh, and chanterelle for £150 when dried.

Despite a rapid growth in the activity, only about 100 tonnes are collected annually in Britain, a tiny figure compared with, for example, the 20,000 tonnes exported by Poland in the first eight months of 2001 alone, and just 10% of what our domestic market consumes (the rest is imported from Eastern Europe). So what does the future hold? Indisputably, edible and non-edible species of fungi are in decline throughout much of Europe, due mainly to pollution and the loss of habitat. As for the effects of collecting *per se*, some people maintain that the persistent removal of the spore-producing parts of fungi may affect their future populations, while others argue, on a different time scale, that taking these parts is no worse for fungi than collecting apples is for an orchard. Whatever the case, scientists certainly need to determine the relative importance to fungi of their spread through mycelium or through spores. Less uncertain is that trampling, and raking of the ground that some collecting requires, may have a direct impact on mycelia.

Many seasoned fungi collectors believe that the collecting is not only sustainable but can even stimulate the production of further fruiting bodies. At the moment, although none of the sought-after edible species are threatened at a national level, some land managers will continue to ban fungi gathering in areas that are more extensively exploited.

In Epping Forest (Essex) the authorities have introduced a licensing scheme that allows local collectors to pick up to three pounds (1.4 kilo) of fungi on one day per year for home-use only. The licenses are free, and are restricted to one per household. In France, regional authorities issue licenses, at £750 apiece, to control professional pickers rather than local amateurs. In the southern département of Lozére, for instance, each person is limited to a maximum of five kilos of fungi per day, and in neighbouring Ardéche to ten. In Britain the activities of professional or semi-professional collectors may upset local people who enjoy seeing the fungal 'bloom' of autumnal woodland floors. But the bigger issue for everyone is the widespread and continuing decline of fungi – edible or not.

Little is known about the true impacts of fungi collecting and in some areas, as here in the New Forest, the activity is banned during certain times of the year.

Marsh samphire is sometimes known as 'poor man's asparagus'.

Marsh samphire

The coastline of North Norfolk and Lincolnshire is the place to see and sample a delicacy known as glasswort (from its former use as a source of alkali for the glass industry) or marsh samphire, locally pronounced 'samfer'. The names in fact cover several similarly fleshy, mainly annual species growing on salt marshes but the main one exploited is *Salicornia europaea*, an early coloniser of bare and frequently tide-covered mud.

In *A cook on the wild side*, Hugh Fearnley-Whittingstall enthusiastically describes marsh samphire as "one of the most distinctive and worthwhile of all the vegetables, with its excellent *al dente* crunch and distinctive sea-fresh salty flavour". Young plants can be eaten raw, but are mostly either boiled or steamed for 5-15 minutes until just tender (they take longer to cook as the season progresses). The flesh can then be stripped from the fibrous stems and is best served simply with butter or vinegar, or to accompany seafood.

Ungracefully slipping and sliding knee-deep in dark, sticky mud is not food gathering at its most glamorous, yet local people have been collecting marsh samphire between June and September for generations, wherever it is common but especially in East Anglia. Here holidaymakers have also joined in to gather it for their own use and, as anyone travelling along the A149 (the Cromer-Hunstanton road) will notice, there is also some commercial activity too.

In North Norfolk the salt marsh habitat of marsh samphire is important for birds.

Norfolk crab and samphire soup
from Cookie's Crab Shop, Salthouse, Norfolk

Ingredients
1 dressed crab
1 medium potato
500 g marsh samphire
600 ml stock (vegetable or light fish)
50 g butter
200 ml single cream
salt and pepper

Peel and finely slice the potato and sauté slowly in butter in a closed saucepan until soft.

Clean and boil the samphire for 5-15 minutes (until the green flesh slides off the stalk). Slide the flesh off the stalk.

Add the samphire to the potatoes and continue to sauté for a couple of minutes. Add the stock and bring to the boil.

Liquidise until smooth.

Return the mixture to the pan and stir in the crabmeat, add the cream and salt and pepper to taste.

Bring to a simmer but not allow to boil.

Serve hot or chilled.

Marsh samphire stalls like this one are typical in North Norfolk.

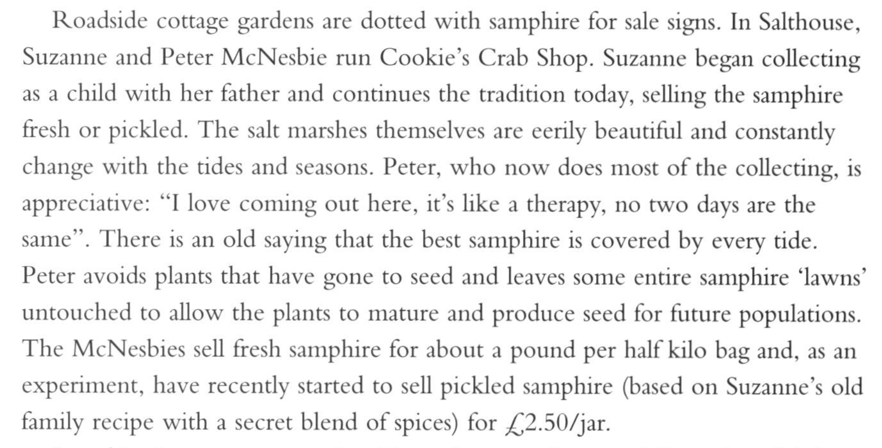

Roadside cottage gardens are dotted with samphire for sale signs. In Salthouse, Suzanne and Peter McNesbie run Cookie's Crab Shop. Suzanne began collecting as a child with her father and continues the tradition today, selling the samphire fresh or pickled. The salt marshes themselves are eerily beautiful and constantly change with the tides and seasons. Peter, who now does most of the collecting, is appreciative: "I love coming out here, it's like a therapy, no two days are the same". There is an old saying that the best samphire is covered by every tide. Peter avoids plants that have gone to seed and leaves some entire samphire 'lawns' untouched to allow the plants to mature and produce seed for future populations. The McNesbies sell fresh samphire for about a pound per half kilo bag and, as an experiment, have recently started to sell pickled samphire (based on Suzanne's old family recipe with a secret blend of spices) for £2.50/jar.

Samphire is strong on novelty. Through promotion as a delicacy by celebrity chefs, its market has expanded away from its traditional locations on the coast to the tables of the smartest city restaurants. But it is not just British plants meeting all this new demand. Rockport Fish Ltd in Tunbridge Wells (Kent), for example, imports all its stock from France, while Wild Harvest, a delicatessen in south west London, gets most of its supplies from France and Saudi Arabia. This imported samphire sells for £4.00-7.50/kilo.

As yet, there is no hard evidence that collecting in Norfolk has dramatically increased, let alone that it is jeopardising samphire populations. However, Peter McNesbie believes that there has been an increase in people indiscriminately collecting to sell at car boot sales. Peter recalls that one of his harvesting areas was completely stripped in two hours.

So, there is clearly some potential conflict between collectors and conservationists. In Britain, plants are usually gathered by uprooting, which is illegal under the 1981 Wildlife and Countryside Act unless the permission of landowners is obtained. In many of the coastal sites where marsh samphire is traditionally collected, conservation organisations are now the landowners or managers. But their main concern is less for the samphire itself than for the effects of collecting it on birdlife: in summer, through disturbance to sensitive breeding species like the redshank, in winter through reduced supplies of samphire seed as food for finches, especially the twite.

So far marsh samphire has not apparently suffered from its increased gourmet appeal. But it would be wise to monitor the situation and, if problems arise, to issue licenses to local collectors. A step in this direction was made in 2001, when English Nature was approached by a food company seeking formal permission to harvest samphire on The Wash. In the meanwhile Peter and Suzanne McNesbie are happy to continue a local tradition and provide a homely service in the summer supplying marsh samphire and the crab and lobster salads for which they are renowned.

Peter and Suzanne McNesbie collecting marsh samphire.

Stinging nettle is one of our most familiar plants. In some countries it is cultivated for its long fibres that are used to make textiles.

Stinging nettle

Whereas marsh samphire clings to the tidal edges of Britain, the nettle (*Urtica dioica*) has settled round the habitations and disturbances of man. Its nasty sting gives it an image problem, yet since Roman times it has been widely used as a food and medicinal plant. Recently restaurant chefs have introduced it into extravagant dishes.

Among the most successful modern adaptations of the nettle as a food comes from Liskeard (Cornwall) where Lynher Dairies wrap it around their Yarg cheese to provide flavour and aid the maturation process. In May and June, and again in September, a small army of pickers, up to 35 strong, peers and stoops among the fields and hedgerows collecting about 2.5 tonnes of leaves – the bigger the better. Back at the dairy, the harvest is weighed, the pickers paid, and the nettles covered with film and deep-frozen until needed. Then, still frozen and in layers, the now stingless leaves are dabbed onto the young, pure white discs of immature cheese.

Collecting from the wild is often labour intensive and expensive. Some attempts have been made to cultivate nettles for use in Yarg cheese, but so far this has proved to be surprisingly difficult as the cultivated nettles have been producing leaves smaller than what is required. Size is not an issue, however, for a new company called Leafcycle, based near Tiverton (Devon), which extracts protein from its own cultivated nettle to produce a vegetable protein it markets as Leafu.

Fruits

Just as spring is the season for greens, the turn of summer and autumn is the season for fruits and berries. Few of us can resist picking juicy blackberries (*Rubus fruticosus*), but the rest of the wild fruit harvest we largely ignore and very little of it reaches any level of commercial exploitation. Where it exists, such collecting tends to be a very localised activity and restricted to just a few people. A good example is the whinberry (or bilberry, *Vaccinium myrtillus*), a common small shrub of heathland and moorland whose edible berries readily turn fingers, lips, and tongue a deep purple. Collecting them used to be a common pastime in many parts of upland or western Britain, but Shropshire is one of the very few places where commercial activity continues. The year 2000 provided a good crop. Two local men sold about 360 kilos of berries to the Stiperstones Inn in Snailbeach where the proprietor, John Sproson, made and sold whinberry pies, crumbles, muffins and sauce. Berries or baked products could also be bought from a local shop called Hignett and Sons Bakery in nearby Pontesbury, and from Harry Tuffins supermarkets as in Churchstoke (Powys). The annual berry crop fluctuates and in 2001 there were scarcely any berries at all.

Rowan jelly, made with rowan berries and crab apples, is delicious served with roast venison. A handful of companies in Britain produces it commercially. As well as rowan jelly, Rosebud Farm in Ripon (Yorkshire) also makes 'wild crab apple jelly'.
PHOTOGRAPH: HEW PRENDERGAST

The fruits of several native and garden escape species are used for preserves and jams. An unquantifiable number of wild blackberry jam jars find their way to many stalls and sales throughout the country. To a lesser extent, the same is true for other fruits such as elderberries, crab apples, rowan berries, and bullace (*Prunus domestica* ssp. *insititia*).

Heather provides a valuable nectar source for honeybees on moorlands and heathlands.
PHOTOGRAPH: TOM COPE

Honey

Honey is produced not by people but by bees. Orchards and crops like oil seed rape (*Brassica napus*) are the source for most British honey, but moorlands and heathlands, where heather is the dominant plant, produce perhaps the most prized honey of all. There is irony in this. Lowland heathland and upland moorland are threatened habitats, and their high value product can provide some small economic reason for conserving them. Heather honey has twice the value of other types of British blossom honey, half a kilo easily fetching a price of £5 or more.

Edible seaweeds

Edible seaweeds cover anything algal from giant brown kelps (*Laminaria* species) to small, red, rock-encrusting species. Not surprisingly, given that the seaweed industry is so competitive internationally, the scale of harvesting in Britain is really quite small. Most seaweed products sold here are imported, like the 'Wild Seaweed Sea Lettuce' (*Ulva lactuca*) and dulse (*Palmaria palmata*) sold by Tesco in 2001 (but withdrawn after low demand). But even so there are a number of people who do make a living from seaweeds in Britain.

In Orkney, for example, the Orkney Tang Company has entered the food business, collecting species such as dulse, tangle (*Laminaria hyperborea*), knotted wrack (*Ascophyllum nodosum*) and carrageen (*Chondrus crispus*) to produce condiments as well as cosmetics, soil improvers and animal feed supplements. Local sheep on North Ronaldsay already eat little else than the seaweeds on the island shores. Diana Drummond, a company in Glen Orchy (Argyll), usually produces skincare products from seaweed, but have recently branched out supplying dried edible species and recipes for using them. Their range includes dulse stir-fry, knotted wrack soufflé and carrageen pudding.

Laver

Many peoples of the Northern Hemisphere have discovered the edible virtues of the membrane–like fronds of red algae from the genus *Porphyra*. In coastal British Columbia and Alaska *P. abbottae* is a major traditional food; in Japan and northern China *P. yezoensis,* known as *nori,* is cultivated on a huge scale; and in Britain we have *P. umbilicalis*. At low tide, draped over the exposed rocks on which it grows, it scarcely looks appealing – and when, 15 hours of boiling later, it has been transformed into a black gel–like mass, its appearance has barely improved. But go

Laver collected from the coastline of Wales awaiting the laborious washing process before it is boiled.

to Swansea, and its three factories processing laver – derived from the Welsh *bara lawr* (*bara* is Welsh for bread, *lawr* means down or floor) – and one learns that for the Welsh, at least, looks are not everything.

Laver is the most important seaweed collected for food in Britain and, once cooked, is known as laverbread. Traditionally it has been enjoyed quite simply, spread on thin toast with oatmeal or malted vinegar and accompanied by fried or grilled cured bacon. More recently, chefs are discovering its culinary assets and serve it with pasta dishes, seafood pizzas and in spicy batter with mushrooms.

Laver is cut mainly by hand at low tide from October to May. In Wales, harvesting is mainly carried out by fishermen and cockle collectors as a sideline to their usual occupation, tending to collect during weekends when they are not fishing. The processing is laborious.

Penclawdd Shellfish Processing is one of the three major laverbread producers in Wales. The company was originally created in early 1995 when a group of cockle collectors invested in a new facility to process both cockles and laver. Success came quickly and in 1998 it was awarded the New Explorer of the Year for their outstanding enterprise.

The collectors bring the laver to the plant in hessian sacks. The first step is to wash it in a series of three steel troughs, each full of fresh water with air bubbling jacuzzi-like through tiny vents on the bottom. Handfuls of the seaweed are tipped into the first trough and the bubbling action washes and floats it towards the end of the trough, where it is picked up by a rotating conveyor belt and transferred to the next one. By the third and final trough, all the seaweed is completely sand-free. The laver is then placed in a steel, gas-powered cauldron with some salt and boiled for up to 15 hours.

The resulting boiled mass is minced and placed in polythene-lined plastic trays for refrigeration and eventually supplied as fresh laverbread to some 40 Tesco stores in Wales and southwest England, and to Swansea Market. Some is canned and sold in shops including Selfridges & Co. in London.

Freshly collected laver at Penclawdd Shellfish Processing is placed into the laver washing machine to remove all sand before boiling.

Once washed, the laver is boiled in water.

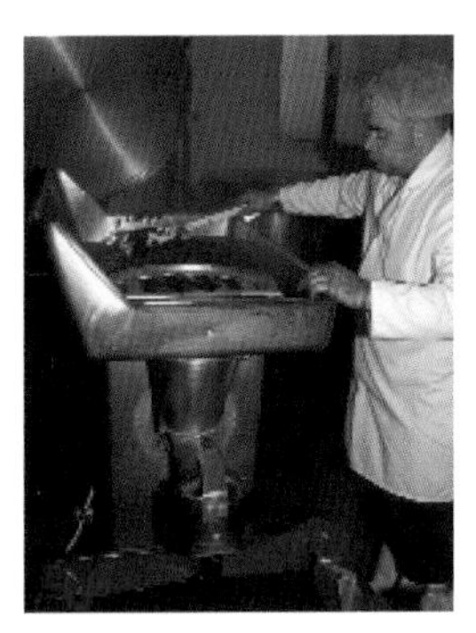

After boiling, the gelatinous black mass is passed through a mincer and is ready for sale.

Laverbread in Tesco in Swansea. Here it is sold fresh by the kilo.

Another of the three processing factories, Selwyn's Penclawdd Seafoods, cleans and processes more than half a tonne a week of laver between October and April (and half this in summer) and supplies laverbread to Asda stores throughout Wales. The third processor is a smaller family-run business called Gower Coast Seafoods.

In 1999, in a pioneering move, these companies subcontracted the canning of laverbread to a Dutch company, making the product a more practical and viable product for home and overseas distribution. Cans of laverbread can now be bought on the Internet.

As well as bordering the Bristol Channel like the southern counties of Wales, Devon and Cornwall also share a tradition for collecting laver. Devon alone has about 15-20 commercial collectors. Among them is Norman Tucker from Appledore who collects from various harvesting sites between Linton and Newquay (Cornwall). He sells unprocessed laver to one of the Swansea processors as well as washed and processed laver to two retail outlets in Barnstaple and one in Appledore. Three other local collectors also supply the Barnstaple outlets. One of them, Massey's Delicatessen, has been stocking laverbread for 18 years, and can sell some 200 kilos a week during the October to April season.

Sensible collecting ensures that the holdfasts (finger-like structures that attach the algae to the rock or ground) of laver are left intact to allow regeneration. Some collectors are concerned that the new demand for laver may encourage inexperienced harvesters to rip up the whole plant, and so destroy a cropable supply for future years. Whatever the future demand actually is – and the Swansea factories report that they are busier now than for many years – the low density of wild laver will always limit supply. It may not be too fanciful to speculate that future supply problems could be solved by any entrepreneur prepared to submit laver to the same cultivation methods already widely used in the Far East for its *nori* relative.

drink

Elder

Elder is one of our most familiar native shrubs or small trees. It is common in most of Britain, although in northern Scotland the introduced red-berried elder (*Sambucus racemosa*) may replace it. After an often spectacular flowering – starting as early as April and peaking in June and July – the elder produces rich purple berries in August and September.

In British folklore and traditional medicine, few plants have featured more than the elder. Its long heritage of use, however, is commercially insignificant compared with the great variety of elderflower drinks stocked by most supermarkets. Yet just twenty years ago, such drinks were no more than the efforts of seasonal family pickings. Elder, a dark horse in the race of wild species towards commercial prominence, has become one of Britain's front-runners.

Three companies dominate the elderflower trade in Britain: Belvoir Fruit

Elder is a plentiful species of hedgerows, woods, waste and rough ground, and on well manured soils.

Farms, Grantham (Lincolnshire), Thorncroft Drinks, Stockton (Cleveland), and Bottle Green Co., Stroud (Gloucestershire).

For Kit and Shireen Morris at Bottle Green it all began in 1989 with a potent mix of entrepreneurial flair and a traditional family recipe for elderflower cordial. They originally sold the cordial through agricultural shows but a lucky break came when it was recommended on the BBC's Food and Drink programme and they began to tout the major supermarkets. They have rapidly expanded and now their factory employs 35 people, produces a whole range of elderflower drinks and other flavours too, and sells them throughout Britain.

To make all their drinks, Bottle Green gets half of what it needs from local hedgerows. In May and June some 600 people are busy collecting the flowers and delivering them to the factory in the afternoon. Many of the pickers devote themselves almost full-time to the harvest for two to three weeks, and the more experienced among them know exactly where to go for the best crop. "People are very secretive about where they pick", says Kit Morris. Restrictions on movement imposed because of Foot-and-Mouth Disease were a problem in 2001 but in 2002 the pickers managed to harvest some 40 tonnes.

Inevitably, with the high demand for elderflowers, producers are now experimenting with cultivation. A couple of miles up the hill from the Bottle Green factory, local farmer Richard Kelly has a plantation of 20,000 elder bushes in a four hectare field. Each year, close to harvesting time in April, he advertises in local shop windows for seasonal pickers, and might employ up to 30 local people at peak times.

A believer in diversification, Richard receives no government subsidy for his plantation. Unsurprisingly, given its abundance as a wild plant, no one has really attempted to cultivate elder for its use in cordial on such a scale before in Britain, so Richard's methods need to be truly trial-and-error. The bushes are pruned to create a bush-shape. They are, however, extremely susceptible to insect damage but, since the plantation is run on organic lines, no pesticides are used. Is the

Elder plantation in Gloucestershire.

Kit Morris holds handfuls of fragrant elderflowers.

cultivation working? It certainly seems so. Not only is the source of the flowers precisely known, but the plantation bushes are already producing as heavy a crop of flowers as hedgerow bushes and, given the concentration of plants, collecting costs should be lower too.

Richard's farm, and another in Kent, between them produce about 20 tonnes of elderflower for Bottle Green. Belvoir Fruit Farms is also pursuing cultivation. It has established 36 hectares of elder plantations to supplement the remainder of its harvest (30-35 tonnes) from local hedgerows and gardens. Throughout Britain a host of smaller companies is also producing elderflower drinks. Rocks County Cordials in Twyford (Berkshire) and Three Choirs Vineyards Ltd, Newent (Gloucestershire) have also experimented successfully with cultivation, but Thorncroft Drinks, Pastures Organic Wine, Tring (Hertfordshire) and Cairn O'Mohr Winery, Errol (Perthshire) still collect from the wild. Lyme Bay Winery near Axminster (Devon) produces elderflower wine using flowers collected from bushes on set-aside land on an organic farm in Wiltshire. One company that acts as an elder fruit and flower supplier for drink producers and herbal medicines in Britain is William Ransom and Sons of Hitchin (Hertfordshire), the UK's oldest independent pharmaceutical company. It employs about a hundred casual pickers to gather about five tonnes of flowers every year and obtains a similar amount from cultivated bushes on two farms in Suffolk.

Once gathered, the flowers must be processed quickly. Bottle Green's warehouse fills with the heady scent of flowers, piled high in plastic crates. The processing begins when the flowers are put into a huge red wine vat and gently rotated and stirred in syrup for 24 hours to extract their delicate summery flavour. The syrup is then finely filtered and used for cordial or other products.

Bottle Green's elderflower cordial.
PHOTOGRAPH: COURTESY OF BOTTLE GREEN

Picking elderflowers from Richard Kelly's plantation.

Commercially produced elderflower drinks have really captured the imagination and the palate. Bottle Green alone is selling over a million bottles of cordial a year plus five to six million bottles of ready-to-drink variations, such as pressé. This success is leading, perhaps inevitably, towards more cultivation although most of Britain's annual 100 tonne crop of flowers is still collected from the wild. While there is still scope for expansion, especially abroad and for the organic market, the one threat looming on the horizon is the increasing use of artificial elder flavourings by certain brands.

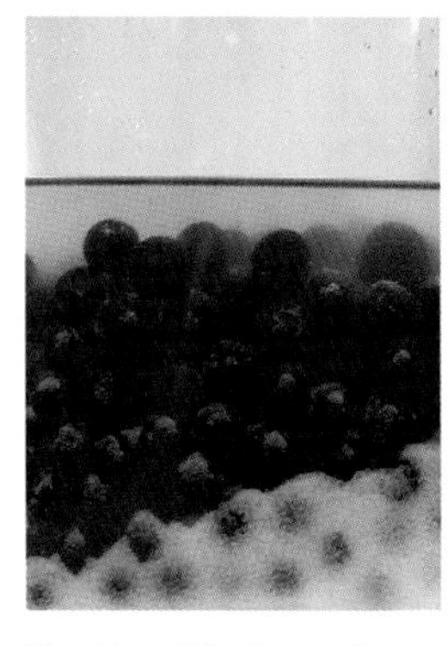

The skins of the frozen sloes split as alcohol and sugar are poured over them.

Sloe

The blackthorn (*Prunus spinosa*) is best known for its crop of tart, acidic fruits – sloes – used to make the deep-red, wintry drink, sloe gin. Though many people make it for home consumption, sloe gin is also produced commercially by a few small enterprises in Britain.

Among them is Bramley and Gage in South Brent (Devon). Although they both come from well-known families of fruit growers, husband and wife team Edward and Penny Kain obtain their sloes from the wild via a team of some 20 collectors originally recruited through an advertisement in a local post office. They include retired people, children, and parents who want to earn some extra cash when their children are at school.

Collecting out in the countryside around Dartmoor and Devon's south coast in October and November, they pick from the hedgerows with the consent of local landowners. Four kilos an hour is a good rate. Altogether, Bramley and Gage use about a tonne a year for their own sloe gin and for an infusion of sloes in cider called Slider; they also supply a further tonne to Plymouth Gin – which obtains the bulk of its sloe supplies from eastern Europe.

The production site at Bramley and Gage resembles a laboratory – white walls, sterile equipment, chemistry-like benches, and a quiet but purposeful atmosphere. The stamp of individuality comes from various customised gadgets, such as pulley systems attached to the high ceiling for straining the fruits. On arrival, the sloes are immediately frozen and, when required, have alcohol poured straight onto them to split the skins (a valuable tip for those making sloe gin at home as it avoids the tedious pin-pricking of each fruit that most recipes call for).

From the hedgerow – sloe gin and blackberry liqueurs.

The end result is bottles of various sizes of Bramley and Gage sloe gin, ranging from miniatures – the biggest sellers of all with 20,000 sold in 2001 – to 70 centilitre bottles. According to Edward "this product sells itself". He and Penny market their sloe gin through various outlets such as farm shops, National Trust and heritage sites, mail order and the Internet. Other companies specialising in sloe gin include Gifford's Hall of Hartest (Suffolk) and the aptly named Sloe Motion near Malton (Yorkshire), which makes sloe chocolates as well. Despite increasing demand, most sloe gin on the British market still uses pulped, imported material but, at some future date, will blackthorn follow the trail of elder and, through its popularity as a source of drink, leave the hedgerows and line up in ranks in farmers' fields?

At Bottle Green masses of flowerheads are stored in crates before steeping in syrup.

A range of cordials from Thorncroft Drinks includes wild-collected nettle, elderflower and rosehip.
PHOTOGRAPH: COURTESY OF THORNCROFT DRINKS

Other species

Perhaps the most northerly, as well as one of the larger drink producers, is Moniack Castle Highland Wineries (Inverness). It specialises in wines, including a pioneering one using the sap of birch trees (*Betula* species). This it collects on or around the Castle grounds as it does for the other wines it makes: the tips of heather, the fruit of rowan and sloe, and the flowers of elder. It also uses the leaves of wild garlic to make a best-selling sauce. In England, one of the higher profile companies making wines from a range of wild plants (among many cultivated ones) is Lurgashall Winery in Petworth (Sussex). It too makes a birch wine, as well as elderflower wine ("hand picked in early summer from surrounding woodland") and liqueurs made from "succulent wild blackberries" and "locally gathered" sloe. Thorncroft Drinks, as well as producing elderflower cordial, collects nettles to make cordial. Cairn O'Mohr Winery gathers approximately five tonnes of young oak leaves from local sites to produce spring oak leaf wine.

Heather Ale Ltd near Strathaven (Strathclyde) specialises in ales based on a variety of plants, most famously perhaps the Fraoch heather ale that is spiced with heather flowers and the leaves of bog myrtle. It also collects bladderwrack (*Fucus vesiculosus*) for its use in Kelpie seaweed ale.

Although the earnings of collectors are often described as being little more than 'pocket money', the costs to companies of collecting wild species are still high compared with using imported material. Yet Britain's wild drink sector is successful and expanding. The most important habitat for its raw materials is the hedgerows that well into the 1990s were still being removed from the countryside at an alarming rate. How apt then that even the commonest shrubs are now producing such delicious results.

Collecting nettles in spring.
PHOTOGRAPH: COURTESY OF THORNCROFT DRINKS

crafts and construction

One of the most heartening features of our countryside is the revival of certain traditional management skills and practices, and increasing public interest in purchasing associated products. Sustainable use of our deciduous woodlands and beds of withy and reed can also benefit wildlife.

Harvesting club rush from the Great Ouse (Bedfordshire).

working the woods

In 1905 England had 220,000 hectares of deciduous woodlands under coppice management. By 1947 this cover had fallen to 134,000 hectares and by 2001 to just 11,600. Scotland and Wales tell a similar tale and now have about 1,000 and 500 hectares respectively.Through the twentieth century, global economics made it cheaper to import timber from other countries. Charcoal and other products lost their markets, people lost their livings and the woodlands fell into neglect. When, in the 1970s, people went back into the woodlands, it was not to seek employment but to conserve the flora and fauna which need the greater diversity of light levels provided by cycles of regular cutting and regrowth.

Since then, the situation has changed again and in some cases these woods can now offer a living. Typical of the several local organisations that have sprung up to help with training and marketing is the Wessex Coppice Group. Its mission statement, unequivocally linking employment and environment, is "To encourage economic growth in the hazel coppice industry through marketing, training and public awareness whilst sustaining the landscape and ecological value of the woodland resource." Wales has its own organisation, Coed Cymru, which provides help and advice on the sensitive and sustainable management of woodlands. Now in Britain there are over 1,000 people coppicing woodlands and selling their associated products.

Charcoal

One product from traditionally managed woodlands that has made a particularly strong comeback is charcoal. Chunks of wood are generally left to dry for 6-12 months, although quality charcoal can also be produced from 'green wood'. When dry, it is split into shorter sections, stacked into a kiln (a technique requiring skill and practice for best results) and partially burned under reduced oxygen levels. Traditionally charcoal was produced in earth kilns, but today the most frequently used modern variations are the protable metal ring kiln and the retort kiln (for smaller producers). The retail price of UK charcoal is about £1.50/kilo, generally 30% more than imported lump wood charcoal. Its selling points include its quick and easy ignition (no need for firelighters or lighter fuel), clean and longer-lasting burn, and benefits to local wildlife through coppice management.

Metal ring kiln in woodland in Cumbria.

As well as with cheap imports, British producers have to compete within a fickle market that is very much dependent on weather and other factors, such as the trend towards using gas barbecues. The purchasing of British charcoal might also stimulate the more sustainable production of the charcoal we import. At national level the BioRegional Development Group, based in Carshalton (Surrey), has played a critical role in its promotion, emphasising the sustainability of its production compared with the less certain origins of imported material which still accounts for 95%, or nearly 60,000 tonnes, of our annual concumption (mostly for barbecues). The British Charcoal and Coppice Specialist Group, a

Ian Taylor adjusts the chimneys
on a metal ring kiln.

This woodland near Witherstack (Cumbria) had not been worked since the First World War until Ian Taylor was contracted to coppice it in 2002. The most frequent species here are hazel and ash (*Fraxinus excelsior*) studded with a few ancient yew (*Taxus baccata*) which will be left alone. These straight hazel poles are used in hedgelaying.

specialist group within the Forestry Stewardship Council (FSC), based in Sheffield (Yorkshire), has developed a scheme to allow small-scale producers throughout England and Wales to sell products from 68 woodlands under the FSC logo which guarantees that the wood is from sustainably managed woodland.

Coppice wood products

Although our woodlands are widespread, they are concentrated in certain regions and the resources they offer may differ: Kent and Sussex, and the North West of England, for example, are rich in woodland products in general, while Hampshire has a high proportion of hazel coppice. Hazel is one of the most important species in our woodlands, and throughout the years of coppice decline it has managed to retain some economic relevance through providing spars to pin down reeds on thatched roofs. The thatching industry still uses some 20 million of them a year, worth some two million pounds. Hazel is also being used, more now than for many years, to make hurdles (for garden fencing rather than the traditional sheep pens) and all kinds of rustic furniture. Other species are turned into tent pegs, tool handles, carvings, and even into an ash used as a glaze by potters. In its garden at Wisley (Surrey), the Royal Horticultural Society routinely stakes up herbaceous perennials with hazel and birch branches. That such a high profile organisation harvests them from the local common is an example of the tide of interest that has turned our attention to the sustainable use of local resources throughout Britain.

Whatever the apparent romance of working in the woods, those who do so do not pretend that earning a living is easy. So what motivates them all? For Ian Taylor the spur was redundancy from an insurance company. After training with a local coppicer, Brian Crawley, he now works in several small woods near Kendal (Cumbria) and produces hazel hurdles, hedging stakes, marquee pegs, birch bird feeders and tables, seats of oak, blocks of wood for turners, firewood and planking. In season, Ian also produces about £200 worth a week of barbecue charcoal.

Hazel poles staked vertically in the ground support this recently laid hedge in Cumbria.

Little known but important coppice products are large pegs used to pin down the ropes of marquees. They are skilfully crafted from a section of ash stem using an axe. The basic shape is first cut from the block, and then the wood is gradually chipped away to produce the peg.

Many coppicers have diversified by producing shiitake mushrooms. To cultivate them, holes are bored into hardwood logs, in this case birch, which are then inoculated with spores and sealed with wax. The logs are stacked in a shady place and the mycelium takes from one to two and a half years to spread.

He sells it with charcoal from other local producers under the label of Lakeland Coppice Products to ironmongers, garages, farm shops, grocers and campsites. Keen to diversify, Ian is also experimenting with the cultivation of shiitake mushrooms (*Lentinula edodes*). This mushroom has been cultivated in the Far East for 600 years, and in the USA since the early 1970s. British grown shiitake currently commands a price of about £6 per kilo, about a pound more than imported material.

Rebecca Oaks, based near Carnforth (Lancashire), came to coppicing after a course in Recreation and Land Management. She exemplifies the hard graft, flexible approach, and marketing and publicity skills required by the successful coppice worker today. She has been working for eight years and, as for other coppicers, her mainstays are charcoal in summer and firewood in winter. During just three weeks in March she can produce 700 bags of charcoal. From April to June each year she also makes about 50 hazel hurdles, sells birch branches, which are used to make shooting hides and show jumping and steeplechase fences, and oak bark to a tannery in Colyton (Devon) for £550 a tonne.

Tapping into interest from the general public, and to supplement her income, Rebecca runs courses and makes presentations at shows. She is also involved in developing longer term support for the coppice industry through her participation in the Bill Hogarth MBE Memorial Apprenticeship Trust, which provides financial support and training for new coppice workers.

As if finding a market for their products was not enough, many coppicers face a new threat from rapidly expanding deer populations. Constant nibbling prevents the straight growth needed for products like hazel hurdles. But there are also new openings and opportunities. Large-scale coppicing could fuel power stations in the future, and there is increasing use of hazel or willow hurdles for riverbank restoration (an example of what is termed bioengineering).

All these activities are providing a new generation of people with a living – and a range of wildlife is thriving too.

Rebecca Oaks splits the stems of hazel before they are woven into hurdles.

Split hazel stems are woven horizontally through vertical stems to produce hurdles of various heights, popular for garden fencing and trellises.

Rebecca supplies birch branches to construct show jumping fences, and makes besom brooms. In its gardens, Kew uses birch besoms for all kinds of sweeping work – from leaves to snow. It purchases 200 besoms/year from a local supplier in Ruislip (Middlesex).

The tanning industry in Chelwood Gate (Sussex) has long gone, but this name is a reminder.
PHOTOGRAPH: HEW PRENDERGAST

Tanning

How do you turn an animal hide into shoe, saddle, purse or any other leather article? Nowadays, this process is done by dunking them into solutions of chromium salts, but the traditional method all over Britain was to exploit the chemical properties of bark stripped off trees, especially oaks. This was done in early summer as the sap began to rise.

In the past, bark was an immensely important part of the rural, woodland economy; now, just one tannery in Britain uses it. On the banks of the River Coly in Colyton (Devon), J. & F.J. Baker & Co. follows a process almost unchanged for nearly 2,000 years; and, as if to prove the point, the business sits atop a Roman tannery. Running the company is Andrew Parr, descendant of one of the Bakers who bought it in 1860. There are 28 employees, most of them from Colyton itself, and two of them with 114 years of service between them.

Just how close the links are between oaks and tanning is shown by the origin of the words themselves. *Chambers 20th Century Dictionary* defines the noun 'tan' as "oak bark or other material used for tanning", and specifies the derivation of the word from the Old French *tan* or the Breton *tann* meaning "oak".

Until at least the middle of the nineteenth century, the Sussex name for the wryneck, a small species of woodpecker, was rinding bird. Its arrival in early summer heralded the right time to fell oak trees and remove their bark – or rind – for local tanneries. Now the county has lost both the bird and the tanning.

Andrew Parr stands next to ground oak bark used to tan leather.

Bark arrives at the tannery in large chunks, and is left to dry in this state before being ground up.

When the bark is ready to be used, it is ground to a finer texture using a grinder, powered by the River Coly, that works as well today as when it was installed in 1881.

Every year J. & F.J. Baker & Co. uses 15 tonnes of dried bark which it obtains from coppiced oak in the Forest of Dean (Gloucestershire), Cumbria (where suppliers include Brian Crawley and Rebecca Oaks) and from several coppice workers in Wales. The bark is peeled from the stems by hand and dried by the coppicers before being sent to the tannery. There it is stored in a huge pile in a barn where it is left to dry for up to three years.

Tanneries are notorious for their foul stench but, in walking around, Andrew Parr promises that every step in the process is progressively better. Both the hides and the bark need preparation before they meet.

Tanning pits where oak bark is steeped in water to produce the deep red-brown tanning liquor that is the vital mainstay of this factory.

First, the hides are salted to dehydrate and preserve them temporarily; in this state they can keep for up to three years. Next they are placed in large liming pits for 14 days to remove the fat and hair. Prior to the outbreak of Bovine Spongiform Encephalopathy (BSE), the fat was rendered down and sold on to produce stearic acid for soap-making, but this is now banned and all by-products have to be incinerated at the company's expense. A resource has become a waste.

The bark meanwhile goes through its own traditional stages. It is first ground down into smaller pieces, and then soaked in water in stone-lined pits for at least a month in order to leach out the tanning liquid whose soft, floating, bacterial crust looks sinisterly medieval. The bark is reused to extract any remaining tannins and finally, as a rich, brown crumbly material, sold as mulch.

After neutralising in a very mild ammonium chloride solution, the hides – 60 at a time – are ready for tanning. Attached to a rope, they are lowered into another series of pits containing the bark liquor and rotated every day to ensure they all receive the same strength of solution. For the first three months they are suspended vertically, and for the next nine months are layered horizontally. As time progresses, the hides are moved to pits with ever stronger tanning solutions.

After tanning, the hides are put through a rolling machine and are dressed in a variety of ways depending on the final use of the leather.

Inside the tannery, the stacked or hanging tanned hides await rolling through a heavy iron wheel to even out their thickness.

Rubbing with cod liver oil smoothes out any unevenness or creases in the hide, and adding mutton tallow produces a dark, greasy leather used for stirrups and harnesses and the best hunting leathers. Sure enough, the final, deeply glowing leather has a pleasant, sweet smell and is ready to be sent on to specialised producers of shoe soles, saddlery and bridlework.

The 2001 Foot-and-Mouth Disease outbreak had a great impact on production and sales: not just because there were no cow hides available for tanning (the carcasses were all burned), but the market for equestrian products suffered from the temporary ban on hunting. The current threat of a permanent ban is of great concern to the tannery, since the riding community is one of its main consumers.

Despite its traditional customer base, the tannery is seeking new markets for its very high quality products such as Japan. In the meanwhile, Andrew Parr and his staff at J. & F.J. Baker & Co. are busy supplying the thousands of customers they already have.

Dyeing

Although most dyers in Britain are hobbyists using cultivated materials, there are a few small businesses producing dyes, coloured yarns and fabrics from British wild plants. In Wales, there are two using locally collected species such as sloe, nettle, dock (*Rumex* species) and lichens to produce dyes for silks and other textiles. Ford Barton near Tiverton (Devon) specialises in Wensleydale sheep and produces spectacularly dyed wool using plants collected from the wild. The final colour of the wool depends on whether the plant material used is dried or fresh, and the length of time it is immersed in the dye. Dogwood (*Cornus sanguinea*) produces pink, blackberries a soft grey, and lichens a grey-green yellow.

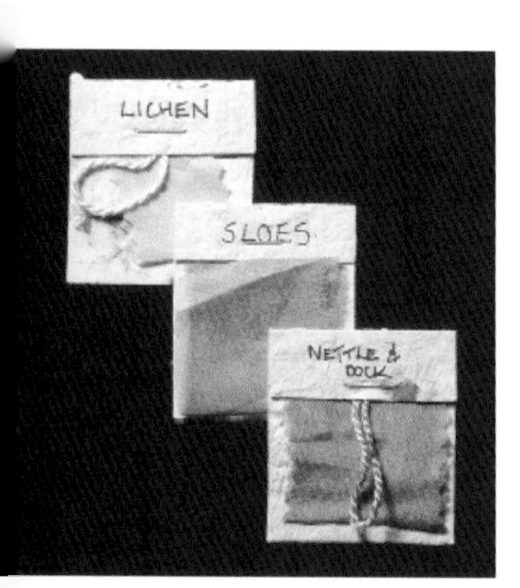

A variety of plants is used by Jacqui Jackson (Wiltshire) to dye silks.

Chestnut-brown hides hanging from hooks await final processing.

roofing

Reed, common in shallow waters of waterways, estuaries, marshes and fens, has a long history of use. The earliest recorded use in Britain dates back to the Mesolithic (about 10,000 years ago) when it was repeatedly burnt to encourage new edible growth for deer on Star Carr (Yorkshire).

Reed

The common reed (*Phragmites australis*) is our tallest native grass and one of the most widespread species in the world. In a happy alliance between conservation interests and a traditional part of the rural economy, its dense stands host some of Britain's rarest wildlife, like the bittern and the swallowtail butterfly, yet also provide the stuff of dreams – the thatched roofs of the country cottage.

Britain has only about 6,500 hectares of reedbeds. Their largest concentration is in the Norfolk and Suffolk Broads where they cover 2,500 hectares, while the single greatest stretch of 800 hectares extends for several miles along the north side of the Tay estuary in Perthshire. Many reedbeds are owned or managed by state bodies such as English Nature or conservation charities like the RSPB and the Wildlife Trusts.

Whether for commerce or conservation, reedbeds need careful control of their water levels. Without flooding, their accumulated leaf litter eventually provides conditions for other plants to move in and replace them. The frequency of reed cutting and removal varies. On most commercial sites there is an annual ('single wale') harvest to produce a dense crop of high quality reeds, but for conservation purposes a series of rotations is preferred with cycles from two years ('double wale') to six years or more. The habits of the reed warbler help explain the difference of approach: only the stems of the previous year's reeds will do for its nest. In some places the two approaches live side by side. The Tayreed Company, for example, harvests part of a reedbed falling within a Site of Special Scientific Interest, and at Abbotsbury (Dorset) contiguous areas of both harvested and unharvested reedbed are subjected to the same summer flooding regime.

As in many other countries, so-called constructed reedbeds have appeared in

Reedbed on Cley marshes (Norfolk).

A reed thatched roof has an expected life of between 50-80 years, as long as the roof ridge is replaced every ten years or so.

The Tayreed Company produces between 100,000-150,000 bundles each year.
PHOTOGRAPHS: HEW PRENDERGAST

Reed is piled high to dry.

Britain as a novel way to treat wastewater from domestic and industrial sources. They may certainly be good for wildlife but it is too early to say whether they will produce suitable thatching material on a commercial scale. A good place to see one is at the London Wetland Centre, Barnes.

An estimated 60,000 buildings in Britain have thatched roofs. Some are listed, many are old, and all will need refurbishment at some time or other. A reed roof made by an expert thatcher will last up to 80 years in eastern counties and 50 years in the wetter south west where a large proportion of England's 600 thatchers is at work. Such is the popularity of thatched roofs that they feature in new housing developments – indeed Dorset County Council actively promotes them – and the thatched cottage dream itself has become an exported commodity.

The Tay estuary has the largest uninterrupted area of reedbed in Britain.
PHOTOGRAPH: HEW PRENDERGAST

On a clear October day Graham Craig, the owner of the Tayreed Company, based in Errol (Perthshire), feeds bundles of last year's reed through a cutting machine to even them all up, while John Shield stacks them onto a trailer in preparation for a long delivery trip. Every winter some of their reed goes to the southern counties of Ireland. A popular retirement or holiday destination for continental Europeans, the area is mushrooming with brand new properties with thatched roofs. Ironically, some of these are more mansion- than cottage-like, and none of them is part of a local tradition.

Tayreed is Britain's largest commercial reedbed enterprise, harvesting up to 240 hectares a year. Every year, starting in December, they cut the reed mechanically and produce two to three thousand bundles a day and up to 150,000 bundles by the end of the season – enough for 50 or so houses. The bundles are piled up into huge ricks on the edge of the reedbed above the reach of the highest tides, ready for further processing at any time of year.

Adding production figures from the Broads – where about 365 hectares are harvested – with those of many smaller areas allows for a rough calculation of 1,000 tonnes a year coming from British reedbeds for commercial purposes. Other material is used directly by producers. At Abbotsbury, for example,

Tithe barn in Abbotsbury thatched with reed from a nearby reedbed.
PHOTOGRAPH: HEW PRENDERGAST

harvested reed is kept for the thatching of a listed tithe barn. In total, though, Britain does not produce nearly enough reed for its own needs. According to HM Customs and Excise statistics, we have been importing nearly 4,000 tonnes a year over the past decade, with Turkey being the main source.

For Graham Craig, with full order books, the future looks good but there are some issues to be resolved. Unexpectedly perhaps, statutory planning requirements regarding re-roofing material on listed properties may be one of them. One authority has refused to allow reed to replace wheat straw (another traditional material) and sanctioned the use instead of a South African substitute. Some people fear the gradual loss of expertise among cutters and poor recruitment because the work is unattractive and lacks any stable employment structure. Others maintain that the trend towards conservation-oriented management of reedbeds produces reed of too low a consistency for thatching.

To meet all its own domestic market needs, Britain would have to increase production five fold, and need an extra 2,400 ha of reedbed to do so. Since increasing the reedbed total in England and Wales alone by 1,200 hectares is part of the national Biodiversity Action Plan anyway, there does seem scope to strengthen the link between the sustainable, commercial use of this habitat and the improvement of the countryside for wildlife. New reedbeds would also create more jobs, directly through their management and through tourism, improve the quality of water passing through them, and reduce the costs and environmental penalties of transporting imports. Increasing the demand for local provenance reed will need education, promotion and certification schemes by companies and planning authorities. Booming bitterns and thatched roofs may have a future in each other's hands.

Saw-sedge

Saw-sedge has a very scattered distribution in Britain. It is most frequently found growing in fenlands and by streams and ponds in East Anglia.
PHOTOGRAPH: TOM COPE

Reed is not our only wild species used on roofs. Because of its flexibility, the saw-sedge (*Cladium mariscus*) has traditionally been used to cover the ridges of thatched roofs and there are still a few places where it is managed and harvested. The most famous site is the National Nature Reserve of Wicken Fen (Cambridgeshire). Drainage has virtually eliminated the fens of East Anglia and Wicken's 370 hectares represent the largest surviving fragment. Ten hectares of it produce a crop of saw-sedge: about 3,000 bundles in 2000 that fetched a pound apiece. The work is carried out as part of the general management, with one person cutting, one or two bundling and stacking, and the others generally clearing up and moving bundles off-site. The small economic return helps offset the costs of conservation.

Harvested saw-sedge in the Cambridgeshire fens.
PHOTOGRAPH: COURTESY OF THE BROADS AUTHORITY

Heather

Britain's moorlands and heathlands once provided an ample supply of heather for roofing the dwellings of local people, but now demand for this use has virtually gone and most of the heather in Britain is unsuitable for thatching having been burned or grazed too short. It produces a dark roof as its County Durham name of 'black thatch' suggests. Rob Littlehales on the Long Mynd (Shropshire) and David Cussons on the North York Moors (Yorkshire) both supply heather to English and Scottish thatchers. The 60-70 cm long bundles fetch about £30 a tonne.

Heather for thatching is harvested mechanically on some parts of the North York Moors.

PHOTOGRAPHS: COURTESY OF THE NORTH YORK MOORS NATIONAL PARK AUTHORITY

Once cut, heather is collected and tied into bales.

Thatcher William Tegetmeier at work with heather in Hutton Le Hole (Yorkshire).

As well as for roofing, heather is also used to make besom brooms in Yorkshire and, in another rather non-traditional way, a company in the Netherlands, Oosterbeek Recycling, packs bundled stems of old, degenerate heather into filters to deodorise air emitted from food factories. Finding a plentiful supply of suitable heather is not easy so, as well as sourcing material from the Long Mynd, it also seeks it from other countries in northern Europe. On the Long Mynd, before the demand for old growth heather for filters was realised, the National Trust who manage the site paid contractors £250 per hectare to cut the plants. Now the Dutch company pays the contractor to provide them with cut materials, saving the Trust over £3,000 a year.

Heather bundled for besom making in Farwath (Yorkshire).

Pollarded willows on the Somerset Levels.

basketry and weaving

As well as ivy (*Hedera helix*), other wild species used for basketry include alder (*Alnus glutinosa*), black poplar (*Populus nigra*), blackthorn, buckthorn (*Rhamnus cathartica*), dogwood, hazel, honeysuckle (*Lonicera periclymenum*), hornbeam (*Carpinus betulus*), field maple (*Acer campestre*), old man's beard (*Clematis vitalba*) and wych elm (*Ulmus glabra*).
PHOTOGRAPH: TOM COPE

All over the world people have woven the stems, roots and leaves of plants into baskets and other containers. It is a skilled and time-consuming activity that in a country like Britain, using plastics more than natural resources, is now more the occupation of hobbyists than professionals. Both groups may belong to societies like the Basket Makers' Association and the Welsh Basket Makers Group. There are, however, more than 300 professional basketmakers in Britain as a whole and, although most use imported materials such as bamboo (grasses) and rattan (climbing palms), more than 80 of them collect a great variety of plant materials from hedgerows and woods. By far the most important plants, however, are willows.

Willow

Few trees are as suited to wet ground as willows – and nowhere in Britain is this clearer than on the Somerset Levels. A unique lowland landscape, crossed by what is claimed to be the oldest wooden trackway in the world (nearly 6,000 years), it is also the heartland of traditional willow cultivation and of an industry which, once central to the local economy, entered decades of decline before a recent and continuing revival.

Based on native species, the willows have been developed into a great number of varieties with curious names such as 'Trustworthy', 'Swallowtail', 'Mealy Top' and 'Dicky Meadows'. They enter this book because, unlike agricultural crops, their management system, location and products have remained essentially traditional – indeed the very presence of the withy beds where the willows grow is a defining feature of the Levels.

Cultivated willow.

Elsewhere, willows are starting to be cultivated as so-called Short Rotation Coppice for biofuel, an exciting development as a source of renewable energy, but one that lacks identifiable links with landscape and wildlife, and so is excluded here. A later version of this book may take a different view.

Willow branches and baskets made by Sheila Wynter, a basketmaker in Stroud (Gloucestershire). Sheila also teaches basketry.
PHOTOGRAPH: COURTESY OF SHEILA WYNTER

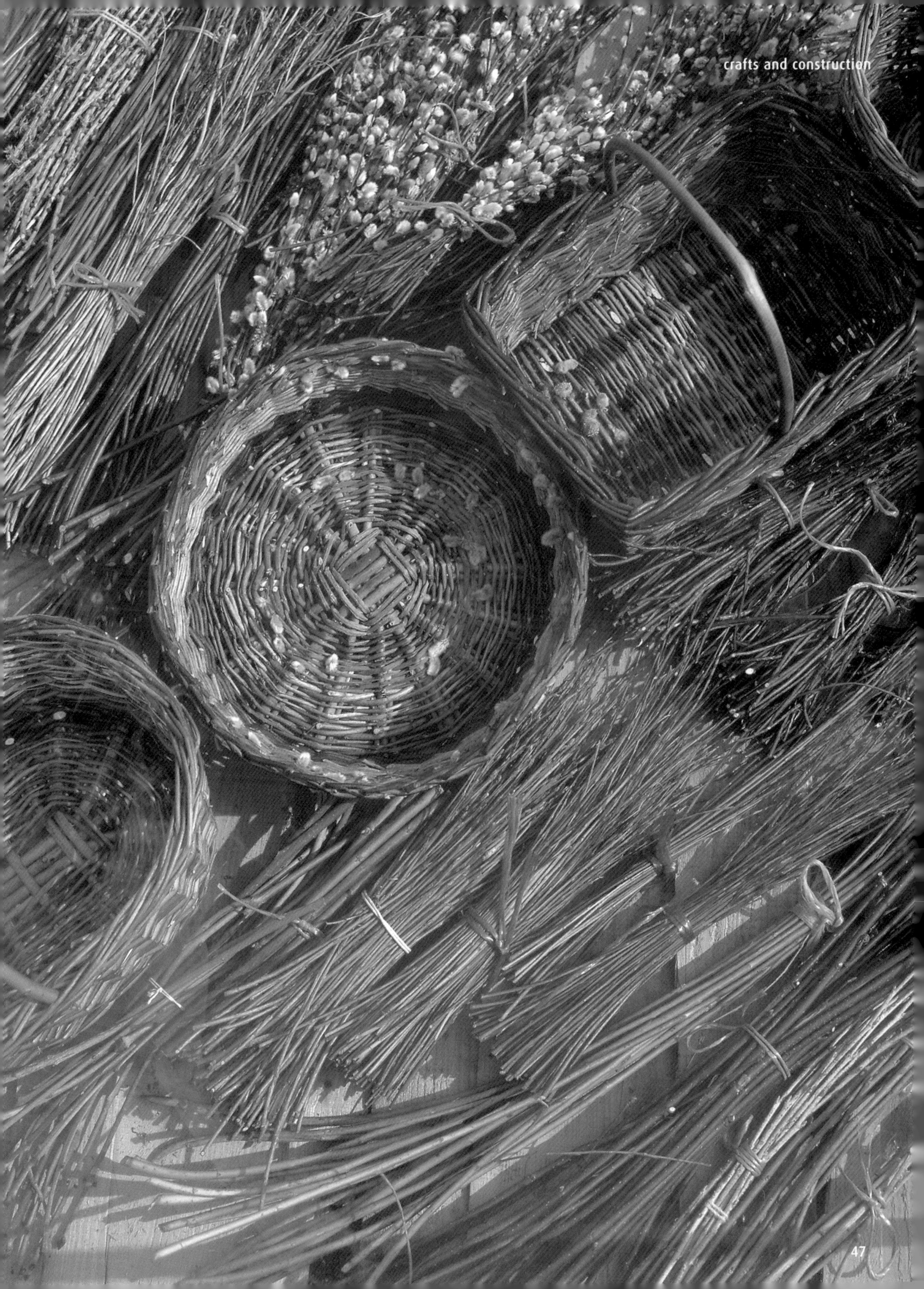

The use of willows for a wide range of farming and fishing equipment – mainly baskets of all shapes and sizes, and traps for eels and salmon in particular – made withy beds a feature of many localities. Even the tiny island of Bardsey (Gwynedd), once said to be home to 20,000 saints, has one. But this use of willow has declined and there are now far from the 7,000 basketmakers there were in Britain in 1945. Even the popularity of the 'crack of leather on willow', as any fan of cricket will know, is not what it was.

The Somerset Levels and Moors, covering about 60,000 hectares, are the largest area of lowland wet grassland and natural floodplain in England. Most of the area lies below the mark of the highest tides, and severe flooding by both sea- and freshwater, and attempts to control or prevent it, have been an integral part of the area's history. Because of its wildlife, notably populations of breeding waders like curlew, lapwing, redshank and snipe, the area has one of the greatest collections anywhere in Britain of national statutory designations, including Environmentally Sensitive Area, Sites of Special Scientific Interest, Scheduled Ancient Monuments, National Nature Reserves, and international Ramsar and European Special Protection Area status.

At P.H. Coate and Son, soaked, softened willow rods are stripped using a machine which uses steel bristles to abrade the bark away.

In 2002 the Levels and Moors became one of twenty-five areas in England with a LEADER+ designation. This is a European Union initiative set up to involve rural communities in testing new ideas for the environmental and economic sustainability of the areas where they live and work. All LEADER+ areas are reliant on a Local Action Group made up of public, private, community and voluntary interests, and in the case of the Levels and Moors, this includes growers, processors and retailers seeking to ensure the sustainability of the willow industry.

It is on the Somerset Levels that both the growing and the processing of willows have survived best. For basketry the most frequently grown species include cultivars of the almond willow, osier (*S. viminalis*), purple willow (*S. purpurea*) and white willow (*S. alba*). Pushed into the soil, small willow branches readily take root and send out shoots. In withy (or osier) beds, cuttings 20-30 centimetres long are used. The growing plants are cut close to the ground annually or biennially to produce a head of straight, unbranched shoots from a single stem. They stop growing at the end of summer, having reached a height of around two metres, and are cut mechanically between November and March. Once established, withy beds may remain productive for 30-50 years.

The Levels have about 135 hectares of withy beds. The largest of the dozen or so growers, with 35 hectares, is P.H. Coate and Son in Stoke St Gregory, which has been involved in the industry for five generations. After 30 years, they replace their beds with grass, and wait a further 15 years before returning the fields to willows.

To escape winter flooding, a specially designed machine (also used for reed cutting) runs on raised tracks to cut the withies. The stems are boiled and stripped of bark (an activity traditionally carried out by local women and children, but now done mechanically), bundled together and finally stacked for storage in a large shed.

Rods of stripped willow are leant upright against a wire fence to dry out.

Dried stripped rods are tied into bundles.

Bundles of willow are stacked in a shed until ready to be woven on site or sold on to other basketmakers.

Making the final product.

When needed for basketry, they are soaked in water to make them pliable. With experience a weaver can produce between three and four complex baskets in a day. Coate and Son uses a quarter of its annual willow production for making baskets on site, and sells a further quarter to other basketmakers.

Deeply rooted though it is in the traditional production of basketry, the business has also diversified into another product line that has achieved a level of global dominance. It makes charcoal sticks that are used by artists and sold in 38 countries. For this two cultivars are preferred, 'Black Maul' (a variety of *S. triandra*) and 'Bowles Hybrid' (developed from *S. viminalis*). Post harvest processing differs from that for basketry. Once stripped of bark and chopped into sections, one-year old withies are placed into large metal boxes and heated without oxygen for nine hours. The result is straight, thin, fragile sticks of various diameters that are then packed into boxes of 25 sticks apiece. Ann Coate says that "you can buy artists' charcoal anywhere in the world and there's a very strong chance it will have come from Somerset." The main competition comes from China. "Artists generally want a high quality product which is why the top brands, such as Pelikan and Fusains, use our charcoal." All export orders exceed 5,000 boxes a time. A niche product, charcoal depends less on visitors and a fickle market than basketry. There can be few better examples of how a traditional activity has adapted to exploit a new marketing opportunity.

From planting to final product, willow is a labour intensive industry. Coate and Son employs roughly one person for every hectare of willow under its control. Eight staff plant, manage, cut and process willow rods, three weave baskets, hurdles and other products, and 17 produce and pack the charcoal. Others work in the company's Willow and Wetland Centre that attracts some 20,000 visitors annually, many of them in school parties. Log and shopping baskets are best-sellers in the shop, while the catering industry puts in orders for products such as cutlery trays.

Short rods of willow that have been cut before packing into steel boxes to be turned into charcoal. These thin pencil-like lengths will reduce in width after the burning due to moisture loss.

Willow sticks before and after burning.

Apart from growers, the Somerset Levels are also home to other commercial workers with willow, among them seven other producers of basketry products (one specialising in air balloon baskets), three sculptors, and makers of furniture, thatching spars, coffins and cricket bats.

The main source, however, not just of British but most of the world's cricket bats, is not here but near Chelmsford (Essex). J.S. Wright and Sons Ltd, having started in 1894, claims to be the "oldest established company supplying English cricket bat willow" (*S. alba* var. *caerulea*). As with Somerset willows, it is planted for the purpose in soils with a high water table, and is harvested usually when 15-20 years old. This specialised trade has evolved a peculiar vocabulary – of sets, pods, clefts and blades – referring to the branches, trees, and stages of the wood from which the bats are made. The blades are graded and sold on to the bat makers who may be anywhere in the world; indeed one of the authors of this book plays with one fashioned in Pakistan.

While British sourced cricket bats – if not British cricket itself – seem to face a secure future, other willow products do not. Home grown ones are competing with imports from Asia and Eastern Europe where labour costs are low. In Scotland, 80% of the baskets sold are imported. Russian willow hurdles are selling for £14, significantly undercutting production costs in Britain.

Yet a brighter prospect for basketry emanates from the interests of people not necessarily professionally involved in basketry. The popularity of courses and a general resurgence of interest in rural issues and traditional crafts may mark a turning point.

A spectacular array of willow bundles and baskets.

Felicity Irons bundles the rushes together.

Club rush

The club rush (*Schoenoplectus lacustris*) is a common plant of still and running waters throughout lowland Britain, although less so in Scotland, the north of England, Wales and Cornwall. When its stems are woven or plaited, it makes a strong material much used in the past for flooring. Now it produces a variety of domestic goods and also plays an unexpected role in the distilling industry.

Britain's largest venture exploiting wild club rush is Rush Matters, based near Bedford (Bedfordshire) and run by Felicity Irons. She employs six people to help with the seasonal harvesting and weaving. Plants are gathered from thick stands in the shallow, slow moving waters of the Great Ouse in Bedfordshire and Cambridgeshire and, in 2001, along the River Avon (Wiltshire) too.

The harvesters manoeuvre their 17 foot long, aluminium flat-bottomed boats through the rushes with a long pole, tugging the blades of their scythes through the stems below the surface, ensuring a clean cut. The stems fall towards them, and are gathered together and their cut ends levelled out against the boat bottoms. Rapid plunging into the water removes them of any dead or inferior pieces. Finally, the stems are tied together with other stems into 'bolts' some half a metre across and piled onto the bow of the boat. Each bolt is enough to cover the seats of about three dining chairs.

It is gruelling work from early morning to evening. On a good day Felicity herself, highly skilled, may collect some 36 bolts and the team as a whole will finish off with more than a tonne. As they work some of the discarded rush clumps together to form floating rafts that snag on the river's vegetation. Felicity

Two or three harvesters work on a stretch of river together, piling bundles on the boat's stern and bow as they go.

At lunchtime, and again at the end of the day, Felicity and her collectors unload their boat of the harvest.

Bundles are stacked onto a trailer for transporting back to the workshop.

Back at the workshop, they are propped up to dry out.

For the construction of some baskets, rush is plaited into lengths, which are then coiled and sewn together.

says that "by morning moorhens will have taken away the remnants to make nests". The harvesters will not visit the same site again for at least three years.

From June to August Rush Matters may gather over 30 tonnes wet weight each month.The company's heavily loaded trailer returns to the workshop, where the deep green bolts are put out to dry. Their colour turns to a rich mossy gold, and the final dry weight for the year's harvest is five to eight tonnes.

With 20% of its annual harvest Rush Matters plaits and weaves mats, baskets, chair seats, bags, cushions, bed heads, screens, lampshades and shoes. The luxury niche market for which these items are intended is reflected in the price – floor matting, particularly popular, sells at £90/square metre.

A stunning collection of products made by Rush Matters.

A further 20% of the harvest goes to 200 other weavers, while all the remainder is sold to the Clyde Cooperage, Lochwinnoch (Renfrewshire) and the Speyside Cooperage in Craigellachie (Moray). There, strips of club rush are inserted as a sealant between oak panels on whisky casks to make them watertight. Both companies purchase club rush only from Rush Matters because it is longer and of better quality than the cultivated Dutch alternative which they use only when they run short.

Although Felicity is finding it increasingly difficult to find suitably large rush populations to harvest, there are no specific conservation issues arising from her harvesting. There is, however, competition from foreign sources. The Waveney Rush Industry in Lowestoft (Suffolk), for example, imports six tonnes a year of cultivated club rush from the Netherlands.

Like the reed, club rush has been planted in water treatment schemes. Whether such stands could also be harvested for weaving is not yet known – but for the time being it will be among the wild, unplanted material that the punts of Rush Matters will continue to glide.

managing the land

From seabed to upland farms comes an array of British species that are used to improve aspects of our environment: the quality of our meadows, soils and water. Their exploitation spans and mixes both traditional systems and the most modern techniques.

4

All but a fraction of Britain's wildflower meadows has disappeared. Now companies are harvesting seeds to recreate such habitats.

Seeds for restoration and gardening

Emorsgate Seeds collects wild seed that is sown to produce a crop of seeds for selling on.

Wildflower meadows

Just as some of the world's rarest animals rely for survival on zoos and captive breeding programmes, many of our own plants also demand a formidable array of skills and approaches to sustain them in the British landscape. Without horticultural expertise the reintroduction of the lady's slipper orchid (*Cypripedium calceolus*) in Yorkshire would have been impossible, and techniques associated more with agriculture than conservation are now being applied at habitat level all over Britain.

The second half of the twentieth century saw all but a fraction of our traditional meadows replaced by improved grassland, arable farming and urbanisation. In the 1970s Dame Miriam Rothschild pioneered the practice of collecting wildflower meadow seeds, and then sowing them on appropriate sites to create or restore wildflower meadows. Now, collecting seeds from remaining meadows has become, perhaps surprisingly, a small but important commercial enterprise, with over 40 companies currently involved. The driving forces have been farmers looking to diversify their sources of income, and organisations seeking to reach conservation restoration targets for our meadows set by local and national Biodiversity Action Plans. The passion for wildflower gardening, and the landscaping of highway verges and developments, have also created markets for seeds.

First established in 1980 and now with six full-time staff and three seasonal collectors, Emorsgate Seeds, based near Kings Lynn (Norfolk), is one of the larger companies specialising in native seed. It first harvests seeds from the wild, noting down the exact grid reference so that their origin can always be traced, then plants them out in 'production meadows' that mature to produce further seeds. Collectors may have to revisit sites in order to catch seeds maturing at different times. The seeds are re-combined afterwards to produce a meadow mix. Cultivated seeds are collected using a brush or suction harvester. Some of these are planted again for further production but most are put into cold storage for later sale. Seeds can survive in cold storage for between five to ten years depending on the species.

Emorsgate markets general purpose meadow mixtures with many wild flower and grass species, such as yellow rattle (*Rhinanthus minor*), common knapweed (*Centaurea nigra*), lady's bedstraw (*Galium verum*), cowslip (*Primula veris*), meadow buttercup (*Ranunculus acris*), field scabious (*Knautia arvensis*) and crested

Richard Brown of Emorsgate Seeds in the cold room where seeds are stored at 4°C.

dogstail (*Cynosaurus cristatus*) as well as specific seed mixtures for other habitats such as wetlands. In total, Emorsgate handles some 200 species.

Although yields of seeds from wildflower meadows are tiny compared with those of cereal fields, they do have an economic value. The Shropshire Wildlife Trust collects up to 75 kilos of seeds per hectare from one of the most productive grassland types in Britain, a neutral pH one classified by ecologists as MG5. After the deduction of costs, every kilo of seeds can yield a profit of about £10.

Dawn Brickwood, Weald Meadows Officer, checking seeds at Church Hill Farm, Sedlescombe, near Battle (Sussex). On one day in August 2002 Dawn and Mick Gosden of Agrifactors harvested 21.5 kg of seeds belonging to tufted vetch (*Vicia cracca*), common bent (*Agrostis capillaris*), ox-eye daisy (*Chrysanthemum leucanthemum*) and red clover (*Trifolium pratense*). By September the seeds had already been used on 11 sites in the High Weald. MG5 meadows contain a minimum of 60 wildflower and grass species compared with an average of just five species in locally 'improved' grassland.
PHOTOGRAPH: HEW PRENDERGAST

A mini combine harvester is used to gather the seeds. The harvested material is sieved to separate the seeds from the tops of the flowers.
PHOTOGRAPH: HEW PRENDERGAST

A British 'hotspot' for MG5 grasslands is the High Weald of Kent and Sussex, where they are the focus of the Weald Meadows Initiative (WMI). This initiative aims "to provide farmers, landowners and other clients with cost-effective, site-specific and practical support to sustainably manage, create and enhance wildflower grassland". The partners and funders of WMI are a potent mix: from Agrifactors, a local company with both the machinery and the expertise to run it, to statutory bodies like English Nature and the Countryside Agency, and charities like WWF and the Denis Currie Charitable Trust. The disappearance of our traditional meadows has a hit a raw nerve. The selling of their seeds, although sounding like a threat, may in fact be their insurance.

The popularity of gardening has no doubt helped to build an appreciation of our native flora. It has also, however, created a demand for an illegal trade in certain species that are propagated not by seed but by bulb division. Particular targets are bluebells (*Hyacinthoides non-scriptus*) and snowdrops (*Galanthus nivalis*) that are sold under the pretence of 'cultivated stock' to unsuspecting nurseries or through popular gardening magazines. This trade echoes the one that uprooted millions of cyclamens (*Cyclamen* species) from countries like Turkey and was stopped only by consumer and conservation pressure in the 1990s.

Trees and shrubs

Even though Britain is one of the least wooded areas in Europe, our most conspicuous plants are still trees and shrubs. The creation of community forests, and increasing interest in landscaping our environment with native rather than exotic species, have produced another set of seed collecting enterprises. The Forestry Commission is a major player, especially in Scotland, its contractors, bands of self-employed seed collectors, gathering tonnes of acorns and huge numbers of seeds of rowan, birch, Scots pine and ash.

Birch produces a vast number of tiny, airborne seeds. Its ability to colonise heathland can be a major conservation issue.

The larger businesses now growing locally collected native species may have a million trees, and spend as much as £100,000 on seeds each year. Among the larger ones in England is Forestart based near Shrewsbury (Shropshire). It employs up to 18 staff and specialises in native broadleaf trees (although it also supplies wildflower meadow and heather seeds). The seeds, either picked directly from trees, or by shaking them and catching the seeds as they fall into a net, are supplied to nurseries and other seed companies throughout Europe and North America.

As with so many other products, cheap imports are a major threat. Hungarian seeds of alder, for example, can be five times cheaper than British ones, but whether they will survive as well only time will tell. Ambiguous or misleading labelling is also an issue. Because no certification is required to prove their exact origin, plants only to have been growing here for six months to be classified as British. The charity Flora locale has introduced a code of conduct to encourage companies to identify the origins of the seeds they sell, and many conservation organisations require ever more local seeds, gathered from close to where they are to be planted.

Seed collecting has become an essential feature of 21st century conservation.

improving the soil

Bracken turning from green to brown in late summer.

Bracken

Bracken (*Pteridium aquilinum*) is one of the most widespread plants in the world and there is certainly no shortage of it in Britain. Here it has had a long history of use in soap making, as a fertiliser (it is rich in potash) and in certain areas it was cut as winter bedding for livestock – a practice that still continues in some places, for example Powys. Today, bracken is a menace on upland farms and moorland, reducing the value of grazing, and is equally invasive on heathlands where it can pose a formidable conservation problem.

In the New Forest the Forestry Commission has historically cut bracken to improve grazing for livestock of local commoners. Piles of the green fronds were normally left to rot but in the early 1990s the Commission began to bag it up for selling to local nurserymen as a versatile horticultural medium and peat alternative. Trials elsewhere have followed suit. On the Long Mynd, the Shropshire Wildlife

The Forestry Commission contract-collects most of the tree seeds gathered in Britain, including acorns.
PHOTOGRAPH: HEW PRENDERGAST

Lakeland Gold bracken compost.

Trust and the National Trust have been assessing whether the costs to farmers of bracken control can be somehow offset. The bracken is cut after the bird-nesting season and is then heaped up for composting. The heat of decomposition breaks down the naturally occurring carcinogenic compound ptaquiloside in the leaves of bracken, so the resulting compost is safe. The product is marketed as Green Frond: 'the environmental answer to bracken control'.

Not everywhere is suitable for cutting bracken. Large areas of the North York Moors National Park, for example, are too rocky for machinery and spraying has been successful enough in reducing the Park's bracken coverage. On their hill farm in Heltondale near Penrith (Cumbria) on the edge of the Lake District National Park, Simon and Jane Bland (trading as Barker & Bland) have developed Lakeland Gold. This mixture of bracken and farmyard manure recreates what was in effect a traditional recipe. The bracken was formerly used for bedding down animals in winter, and the following spring was spread on the land to fertilise the fields. Lakeland Gold comes close to being the perfect countryside product: it exploits a resource that is not only hugely abundant but has become a serious ecological and economic problem.

Money saved can be as important as money made. On the heathlands of Ashdown Forest (Sussex) a major management aim of the Conservators is to reduce the coverage of bracken. Using specialist machinery, staff of Kew's nearby country garden at Wakehurst Place scrape bracken-dominated areas to a depth that removes bracken rhizomes and has enough soil and organic matter to make an acidic mulch for rhododendrons and azaleas (*Rhododendron* species). The scraped area is then seeded with heather cuttings and within three to four years the Forest has gained a patch of high conservation value heathland. Kew gains more than just a huge pile of compost. Its horticultural prominence and educational remit combine to help spread a strong message to Britain's gardeners that there are viable and environmentally friendly alternatives to peat.

Seaweed

Traditionally every community had to make the most of its local resources. In coastal areas an important source of nutrients for the soil was seaweeds washed up on the shore. Now, however, the main species used are the calcareous red seaweeds *Phymatolithon calcareum*, *Lithothamnium coralloides* and *L. glaciale*, referred to collectively as maërl. They are slow-growing, bottom-dwelling organisms and over long periods their dead calcareous skeletons can accumulate into deep deposits overlain by a thin layer of pink, living material. Maërl beds are an important habitat and are particularly well developed around the Scottish islands, in sea loch narrows, and in the Fal estuary (Cornwall).

It is in the Fal estuary that a company commercially dredging maërl is based. The Cornish Calcified Seaweed Co. Ltd (Truro) started in the 1970s, picking up a local tradition that had flourished in the area from the 1750s onwards. The company's dredger, operating out of Newham on the River Truro, sucks up

Maërl, dredged off the coast of Cornwall, is sold as a soil improver and also animal feed supplement.

about 30,000 tonnes a year. This sounds a lot but is tiny compared with quantities collected in France, the source of most maërl fertiliser sold in Britain. Cornish Calcified Seaweed markets its maërl as Calseamin and distributes it through about 30 sales agents and directly to the livestock sector. Full of essential elements for plants, and applied at the company's recommended rate of five hundredweight per acre, the benefits are clear, according to the company's marketing pitch: "enrich the soil – improve grazing – improve fertility – improve health, yield and growth – build bone!" A bulk load of Calseamin costs about £40 a tonne. Due to their fragility, maërl beds are easily damaged and they are the subject of studies looking at the impacts of fishing and other anthropogenic factors.

Another company adapting a local tradition, this time the practice of spreading seaweed on the land, is the Orkney Seaweed Company Ltd. At its head office and production facilities on the northern Orkney island of Westray, it produces a range of fertiliser products based on liquid extraction from freshly harvested tangle. The company is targeting both conventional and organic horticulture and agriculture, and hopes to expand into the sports turf and glasshouse crops markets. At Lunnaness (Shetland) the focus of Böd Ayre Products is knotted wrack, from which the company makes a Liquid Seaweed Extract that is a natural feed for plants, crops, animals and animal feed supplement.

water treatment

As well as reed, bulrush or reedmace (*Typha latifolia*) is one of the most frequent components of vegetative water treatment systems. It can tolerate and accumulate heavy metals such as lead, zinc, nickel, copper, iron and manganese.
PHOTOGRAPH: HEW PRENDERGAST

The conservation value of reedbeds, and the economic value of the thatching material that comes from them, are well known (and covered in chapter 3). In Britain, as in many other countries, reedbeds are also being created for the specific purpose of water treatment from domestic sewage to chemical outflows. A number of companies design and create ecologically friendly water and sewage treatment for domestic use. Most of the approximately 600 so-called constructed reedbeds are very small, serving only single households; the 250 sites of Severn Trent Water are generally between 200 square metres to one hectare in size. There is clearly much scope in using this natural method of treating water. Whether wildlife and thatchers will find them equally attractive only time will tell!

healthcare

Some 80% of the world's population still looks straight to plants for their health and well-being. To protect them and ensure a continued supply, people have evolved their own taboos, traditions and practices. In Britain such a close connection between people and plants has largely gone but a few enterprises provide a fascinating glimpse of the past.

5

Soaps made by Linda Hambler in Kent contain both native and non-native plants.

medicines and plant extracts

Medicines

From Chinese to Amazonian and African, medical systems the world over have plants at their core. Even the most ubiquitous synthetic drug of all, aspirin, has its own botanical ancestors – salicylic acid – in the stem of our native meadowsweet (*Filipendula ulmaria*) and the bark of willows. The revival of interest in herbal remedies has meant that Britain is now one of the world's biggest importers of herbs, and healthcare companies are pouring out a stream of new 'natural' products to meet demand. But despite this interest, our own wild species play a remarkably small role in this market. Almost all the tinctures, creams or infusions we use derive from plants that we import or cultivate.

Christine Herbert, a herbalist practising in Wymondham (Norfolk), is an exception. Although most of the plants she uses she grows on her smallholding, she also gathers some wild nettle, chickweed (*Stellaria media*), cleavers (*Galium aparine*), elder, comfrey (*Symphytum officinale*) and meadowsweet from local hedgerows and lanes. Different plant parts are harvested at different times, for example bark of some species before their leaves emerge, and roots of others after they have flowered and seeded.

Christine Herbert collecting meadowsweet.

Nettle makes an excellent tonic that contains calcium, iron, vitamins and other minerals. Elderberry and elderflower have antiviral properties and are used in expectorants and to increase perspiration. Elderflower is also used against hayfever together with plantain, and with nettle and eyebright (*Euphrasia* species) as an infusion and eyewash. Willow bark is used for arthritis and pain relief, and astringent oak bark Christine mixes with abrasive field horsetail (*Equisetum arvense*) to make a powdered toothpaste.

From all her collections Christine makes 150 different tinctures as well as herbal teas, infusions, powders, ointments and oils. Business is good. New patients appear every week, and for all of them she tailors prescriptions made from a combination of up to five different plants.

Other companies include the Orkney Tang Company, which markets a range of veterinary products using tangle, and Atlantic Resource Development (South Uist), which sells the dried peeled stems of this species to a Swedish manufacturer for the production of cervical dilators. The dried stems are sold for approximately £3,000 a tonne.

Extracts

A somewhat celebrated story of the 1990s was the sudden rise to prominence of the anti-cancer drug taxol, originally sourced from the foliage of the Pacific yew (*Taxus brevifolia*). Several small enterprises around Britain sprang up to collect clippings from our native yew (*T. baccata*) to sell them on to European companies.

Friendship Estates Ltd near Doncaster (Yorkshire) exported 15 tonnes of dried, chopped material in 2001 to Germany. Most of the yew was actually not truly wild, but from hedges and topiary in private gardens and estates. The future, however, is uncertain, as cultivated yew will become a major source of the raw material, and synthetic taxol may even eventually replace the natural source.

Brown seaweeds (Phaeophyta) contain alginates, polysaccharides that have a range of applications in food, paper, textile, pharmaceutical and other industries. The only alginate processing factory currently in operation in Britain today is ISP Alginates in Girvan (Strathclyde), the world's leading supplier. First established under different ownership in the 1930s, the factory used to be supplied with tens of thousands of tonnes of seaweed, particularly of tangle and knotted wrack from crofters of the Western Isles and Orkney.

Today, more than 10,000 tonnes (dry weight) of exotic seaweeds are processed at the factory, most of which is imported from Australia, South America, Iceland, Ireland and South Africa. Only about 500 tonnes (dry weight) each year is of British provenance, supplied by some 20 crofters in the Western Isles. Tangle collectors will generally sell seaweed dry, and hope to receive nearly £200 a tonne. Knotted wrack collectors will generally sell their product wet for £20 a tonne. Crofters traditionally make a living from a variety of seasonal activities, and seaweed once contributed a significant proportion to their income, so a downturn in this market has been a major setback. The Gaelic-speaking crofters also collect tangle and knotted wrack washed up on the beaches in winter and supply them to a few other companies who use them for fertilisers and animal feed.

Christine's herbal pharmacy.

products for the skin

Read the label of almost any bottle of shampoo or body lotion and it will invariably contain plant-derived ingredients, yet most plants used in toiletries and skincare products in Britain are sourced overseas. The demand for natural, organic or fresh ingredients is on the increase, however, and there are about a dozen small but thriving companies using wild native species to make shampoos, creams, soaps, face masks and other skincare products.

A mixture of cultivated and wild plants is at the heart of Merrywood, the name for Linda Hamblen's soap making company based in Mereworth (Kent). The wild plants she collects from hedgerows and woodlands are scarcely exotic: nettle, elder, roses (*Rosa* species), comfrey, alder and beech to name a few. But the soaps she makes from them are popular luxuries, with all the modern allure of natural products.

Because of their high mucilage (a gelatinous substance used to moisturise, soften and soothe) and mineral content, seaweeds too are important and frequent ingredients in shampoos, face and body masks, creams and lotions. Few of them are collected from Britain, however. One British company that uses local seaweed is Diana Drummond, run by John Kerr and his wife Erica based in the Bridge of Orchy (Argyll).

The company was set up in the 1950s by a farmer who discovered the valuable properties of seaweeds, which he used for animal feed, after they healed a skin problem on his wife's hands and arms. He developed several recipes for lotions and creams, but after his death in the 1970s the company ceased trading. In 1991 John and Erica, two original customers of Diana Drummond, revived the

Linda Hamblen collects, dries and grinds rosehips before adding then to the soaps.

Soap is initially moulded into large blocks that are wrapped and stored in wooden drawers. Individual soaps are cut from this into 100 gram sections and are sold for £2 a bar.

Merrywood makes some 1,500 bars a week, and sells them at craft fairs and over the Internet.

Rock pools contain a diversity of species including toothed wrack (*Fucus serratus*), sea thong (*Himanthalia elongata*), pepperdulse (*Osmundea osmunda*) and coral moss (*Corallina officinalis*).

company with a whole new set of tried and tested recipes and today have 2,000 customers worldwide, trading via shows, mail order and the Internet.

Gathering the seaweeds is expeditionary. In a small, adapted crabbing boat John and Erica set off for remote rocky islets off the Isle of Mull, often passing whales and dolphins on the way. From April or May to September, needing both low tide and good weather, they harvest their raw material from among a wealth of richly coloured species semi-submerged and draped over the rocks, or waving back and forth in the crystal clear water.

John points out that while all the species are edible and rich in minerals – red ones are good sources of iron, green ones contain lots of calcium – the main

On rocky islets off Mull the low tide exposes the drooping stipes of kelp (*Laminaria digitata*).

John and Erica Kerr anchor their small boat and paddle ashore in an inflatable.

Kelp stranded on the rocks at low tide is used as hand grips for John and Erica to pull themselves up from the slippery intertidal zones.

John and Erica Kerr carefully cut the green fronds of carrageen from the holdfasts.

species they are searching for are carrageen and knotted wrack. Their annual harvest of carrageen is only about 50 kilos dry weight – but more than ten times wet weight is cut by scissors and hauled into the boat. They visit sites on a rotational basis, not more than once a year, to avoid over-collecting.

The gathered seaweeds are dried in the sun during the summer, and are then processed in different ways. Knotted wrack can be finely powdered and added to various products while carrageen is most frequently boiled to produce a gel-like substance. This alone is a superb emollient for the skin. Diana Drummond's other products include cleansing and moisturising creams, face masks, bath soaks, soaps, hair shampoos and conditioners. More recently the company has branched out into marketing packs of edible species, and offering recipes like carrageen pudding, knotted wrack soufflé and a dulse stir-fry.

Despite the massive trade in natural products for use in healthcare, Britain's wild species currently play just a small role. The constraint may simply be one of economics: our labour costs far exceed those of most other countries, especially Eastern Europe where many herbs originate, and home-grown companies can only compete where there is a high degree of value-added to the raw plant products.

Knotted wrack washed up on the rocks.

Kelp.

decoration

An urban society like Britain's expresses a need for nature by bringing plants indoors. We use a surprising array of species – from the lowliest to the loftiest.

A handful of little shaggy moss (*Rhytidiadelphus loreus*) destined for the floristry market.

foliage

Daisies are at their best *en masse*.
PHOTOGRAPH: HEW PRENDERGAST

A huge variety of plants decorates Britain's houses and work places – and some even adorn us in the form of jewellery. Heathergems of Pitlochry (Perthshire) uses the stems of heather cut from the surrounding hills. After drying and removal of their bark, the stems are dyed different colours before being pressed into resin blocks under high pressure. These blocks are sliced and shaped and finally lacquered to make brooches, rings, pendants and bracelets. On a smaller scale, Daniel Treger in St Leonards-on-Sea (Sussex) weaves rings and brooches from the stems of species such as ivy and plantain which he sells at craft fairs. The popular pastime of pressing flowers and leaves has also been taken to a commercial level. One of possibly many small enterprises doing this is Sarah Feather Design Ltd, based in London, that makes framed pictures, cards and lampshades. Among the species it uses must be one of the most lowly, yet most familiar and beloved of all, the common daisy (*Bellis perennis*).

These uses, however, are minor on a national scale. While special occasions and simply a yearning for colour provide a huge market for cut flowers now flown in from all over the world, our greatest festival still relies on a triumvirate of common, native, evergreen plants: mistletoe (*Viscum album*), holly (*Ilex aquifolium*) and ivy. No Christmas would be complete without them.

Mistletoe clumps in a leafless winter apple tree, Herefordshire.

Mistletoe

Most parasites fill people with horror but not the mistletoe. The endorsement to kiss beneath a suspended sprig represents the very spirit of contemporary Christmas, despite the Druidic origins of the custom. As with the holly, the greenery of the mistletoe's leaves, and the abundance of its berries in the depths of winter, are obvious signs of vitality at a time when much else is dormant.

In Britain the mistletoe is partial to growing on trees in the rose family. Common hosts include the hawthorn, blackthorn and rowan but the favourite one is apple and where there are plenty of orchards there is plenty of mistletoe too. It is no surprise then that the commercial centre for the plant is Tenbury Wells (Worcestershire), affectionately known by Queen Victoria as the 'Town in the Orchard' since it lies in the heart of the Vale of Evesham, one of Britain's great fruit-growing centres.

Once a week for three weeks before Christmas, the yard of Brightwells Auctioneers comes alive with the only specialist mistletoe auction in the country. Row upon row is laid out on the ground, trussed up with string and sold one lot at a time. Apart from orchard owners grateful for a Christmas bonus, the main suppliers to the auction are Roma people. According to Bobby, a Roma who lives in Tenbury Wells and whose family and friends are involved, hundreds of people gather the plants from surrounding counties in early December. They seek permission to collect and appreciate that the highest prices are fetched for bundles of mistletoe free of the branches to which they were once attached. Careful pruning allows the mistletoe to regenerate.

Prospective buyers at the Tenbury Wells auction.

Once the auction is underway, auctioneer Nick Champion runs through the lots quickly, followed by a gathering of eager buyers and sellers. As each lot is sold, John Hill rings a bell to signify the fact and to move everyone on to the next one.

The year 2002 was a good one for the collectors as prices were high – a nine kilogram lot fetched £20-£30. Some buyers chop their lots into smaller bunches and wrap them in cellophane for sale in supermarkets. There they fetch about £1.50 apiece, particularly in London. Every Christmas, Brightwells auctions some 15-20 tonnes of mistletoe, so total sales may reach some £60,000.

John Hill holding a bundle of mistletoe in each arm. He has been employed at the auction all of his working life.

The biggest threat to mistletoe comes not from collecting – despite the destructive tendencies of some 'cowboy' collectors – but from the continued grubbing up of old orchards. It would be ironic if some of them could be saved by income from the plant that parasites them.

Holly

Shiny, prickly leaves and brilliantly red clusters of berries make holly one of the easiest British plants to recognise. Add this to its wide distribution and it is easy to understand why it is the most abundantly collected Christmas decoration. Most of what is traded can be found alongside the mistletoe at Tenbury Wells.

The Roma have a rich knowledge of plants and their uses. Bobby, for example, explains how they used to make a glue from holly bark to catch birds for caging. Everywhere in the world such traditional knowledge is disappearing and there are concerted efforts to record it. Ironically, British researchers have done more studies on peoples of the remotest parts of Africa, Asia and the Americas than on the inhabitants of our own islands.

Holly and mistletoe lots await viewing.

Scots pine outside the Ashdown Forest Centre. Sections of birch trunk provide a secure base for the trees.
PHOTOGRAPH: COURTESY OF CONSERVATORS OF ASHDOWN FOREST

The last decade, however, has seen a turn in the tide. The Flora Britannica project revealed the extent to which plants still introduce meaning into our lives and culture. The Flora Celtica project, coordinated at the Royal Botanic Garden, Edinburgh, covered the traditional uses of plants in Scotland and explored their economic potential. But more needs to be done to halt the loss of our traditional knowledge about plants – part, after all, of our cultural heritage.

Farmers require Bobby and other holly collectors to cut back trees or bushes so that they retain their overall shape. Once at auction the holly lots fetch £30-£40 apiece, the highest prices going to those containing cultivated, green-gold variegated types with plenty of berries. One holly buyer, Peter Summers, attends the auction in order to make Christmas wreaths. He used to make about a thousand every year but now makes less, despite the demand being as high as it was 20 years ago. Another buyer is a florist who has a commission to make 400 wreaths during the run-up to Christmas. Holly wreaths sell for prices ranging from £5.50 in markets to £9.50 in florists and garden centres. Wreath makers also buy small amounts of the ivy on sale.

Christmas trees

On Ashdown Forest (Sussex) Scots pine has been gradually establishing itself as a Christmas tree. It is a rapid coloniser of the Forest's important heathlands and in December is removed as part of routine conservation work and put up for sale at £1.50 a foot. In both 1999 and 2000, sales from some 400-500 trees amounted to £4,500; since then supplies have dried up, literally, because of fires that have swept across hundreds of hectares. Although the Scots pine does not compare in quality with traditional species of spruce or fir (*Abies*), customers are content about the lower price and the fact that they are supporting heathland conservation. A problem plant has become a source of income.

moss

Evergreen species are not the only ones entering commerce for decorative purposes. One of the larger companies collecting other plants is McPherson Atlantic. It collects foliage from a range of species including bog myrtle, willow and hazel, heather and rhododendron (*Rhododendron ponticum*). John Verbeeten of Colgate (Sussex) supplies over 70 florists with the stems and foliage of various trees and shrubs and herbs like teasel (*Dipsacus fullonum*) and golden rod (*Solidago virgaurea*).

The scale of all this collecting for decorative purposes is really very small at national level and of little concern, but for bryophytes a different picture is emerging. Britain is extraordinarily rich in mosses and liverworts (there are about

The brush-like heads of teasel, as the name suggests, were once widely used in Britain's textile industry to tease out wool fibres before spinning.
PHOTOGRAPH: TOM COPE

1,000 species), especially in the wetter north and west, although the notion that they have an economic value might seem surprising. But look into any florist or garden centre and there in the hanging baskets and around the bases of pot plants is a thick green – and attractive – carpet of them. Where do they come from?

Much British material comes from land owned by the Forestry Commission and, in Wales, Dŵr Cymru (Welsh Water). A number of small enterprises is involved which require licenses for their activities. The main species are heath plait moss (*Hypnum jutlandicum*), little shaggy moss, red-stemmed feather moss (*Pleurozium schreberi*), glittering wood moss (*Hylocomium splendens*) and several species of *Sphagnum*. The lichen *Cladonia portentosa* is also collected (the closely related *C. stellaris* is imported to make 'trees' for model railways).

Among moss collectors in Wales is John Spikes, who lives just outside Aberystwyth (Ceredigion). As well as collecting from his own 32 hectare forest, he also operates in local conifer plantations, often with his employee, Reuben Owen. They collect moss by hand. Raking mosses, says John, trebles the time they take to grow back. Sites are harvested every three to four years.

A similar rotational system, with concern for sustainability, is also at the heart of the collecting operations of the Welsh-based company Booth Moss and Foliage (Colwyn Bay, Conwy) which, as well as working in Wales, collects from Scotland too. It regularly harvests in forests about to be clear-felled, the main months of work being February-May and again in October. Every year the company collects enough moss to fill 50 articulated lorries.

The velvety green mossy ground cover below a Sitka spruce plantation near Aberystwyth.

Little shaggy moss.

Reuben Owen collects moss from beneath a Sitka spruce plantation.

The moss is packed into bags.

There are two distinct markets for the moss: in winter it goes to wreath makers, and in summer for hanging baskets. John Spikes sells most of his stock to about 20 florists in south east England, generating some £16,000 annually. He believes that there is a huge potential for non-timber forest products (such as mosses) in British forests, especially as timber itself may be of little value. Operations must, however, be small to be sustainable. He also believes that, in order to be healthy, woodlands must be used.

Nonetheless, moss collecting in Britain is raising disquiet in some quarters. Even with licensed collectors it is difficult to establish which species they target and how much they actually gather, and information on harvesting and regrowth is severely lacking. There is also a problem of illicit gatherers taking moss from unlicensed sites. In plantations about to be felled, however, where the resulting change in conditions will adversely affect mosses, there does seem to be some opportunity to expand activities. As we have seen elsewhere, research is required to establish the 'ground rules' for sustainable collecting.

Bags of moss are sold for £4.50 each to the hanging basket trade.

sustainable uses and the future

As our television screens bring us pictures of forest destruction in the Amazon and elsewhere, it may be difficult to equate chopping down our own trees with conservation and sustainable livelihoods. By getting the balance right between what grows and what we take, people, plants and the places they live in can all benefit.

Bluebells thrive in coppiced woodland.

Making charcoal at Wakehurst Place.

At the start of the 21st century few people might expect wild species to offer much employment in Britain. Our country is, after all, one of the most developed and highly populated parts of the globe, with an industrial history second to none. The fact remains, however, that some livelihoods are stronger now than they have been for generations. Conservation interests, rekindling traditional skills and crafts, have returned many woodlands to long lost management regimes; the demand for thatched housing has rescued many of our reedbeds from economic moribundity; and some current activities – like the harvesting of wild flower seeds for meadow restoration – scarcely even existed little more than a decade ago. And who would have guessed that the British would be persuaded to overcome their ancient distrust of fungi or that elderflower drinks would have secured their place on our supermarket shelves?

Meanwhile, scientific scrutiny of the economic potential of our native flora has continued apace and sprung its own surprises. In the 1980s and 1990s pharmaceutical companies combed the world's tropical forests for novel compounds, often relying on local knowledge about medicinal plants to guide them. Now, with the growing realisation that maybe our own flora and folklore have as much to offer, Abernethy, Anglesey and Aldershot may be as much under investigation as Africa and the Amazon. Industry, academia and farmers are exploring the possibilities of growing new crops in the British countryside. Fields of cultivated medicinal dandelion (*Taraxacum officinale*) now adorn parts of East Anglia and the thrust for sustainability will no doubt bring many other of our native species into the laboratory and then into cultivation. Boundaries between what we call 'wild' and 'not wild' are blurring for more than just species. The wildflower meadow habitats made by conservation organisations, for example, are increasingly difficult to separate from the creations of wildflower gardeners.

The people concerned have changed too. As well as bearded septuagenarian or lonesome outdoorsman, today's charcoal burner is just as likely to be female as male, and to be a college graduate of some countryside management course. The overwhelming majority of the people we have encountered in our research are enthusiastic, highly dedicated to their profession and committed to the environment. The wild species from which they earn a living stimulate and tax their entrepreneurial spirit and force them to seek opportunities to create and find new markets. Country fairs, now abundant and popular, provide the perfect venues to educate the public about the products and the benefits they give to the habitats from which they come. The Dorset Coppice Group (formed by local members of the Wessex Coppice Group in 1999) has a typically direct message: "A wood that pays is a wood that stays". Such an organisation can coordinate and promote the efforts of those who (in its own words) "tend to lead quite isolated lives, often tucked away deep in the woods with few people knowing of their existence".

But just how sustainable is the current use of our wild species? Although it might seem naïve simply to classify all traditionally-based activities as such, there

Some of our most treasured landscapes, like the Lake District, have evolved through centuries of human interaction with the natural world.

is much evidence to support the notion. Many individual broadleaved trees, and shrubs like hazel, have had their lives extended for centuries, even millennia, by a continuous cropping of their stems. The woods themselves have ancient histories of exploitation.

For other species there is far less certainty. For fungi populations, scientific knowledge about the long-term effects of harvesting is absent, although debate is rife. About mosses too there is little consensus. In some cases, the sustainability issue surrounds not the species on which livelihoods depend but the economic viability of the whole activity. Bracken that is collected, processed and sold by farmers must compete against well-established horticultural mulches and composts, while the future of British products like hazel hurdles, willow basketry and thatching reed is threatened by cheaper imports, a problem that might get worse with EU enlargement.

Wild species will always be tempting to those who see abundance and easy access as a path to fortune. Instead, it can lead to court, as in recent cases of removing mosses from Hankley Common, a Site of Special Scientific Interest in Surrey, and uprooting bluebells to sell to garden centres. In some areas, bans have been imposed on fungi collecting due to the insensitive large-scale activities of certain individuals.

There is also the potential for conflict between wild species, harvesters and landowners. In North Norfolk the conservation bodies that own or manage much of the coastline have so far tolerated the traditional and relatively low levels of samphire collection by local people. But what happens if gastronomic demand rockets and the breeding salt marsh birds become so disturbed that they start to move away? Will an outright ban be imposed or will there be moves towards the creation of new salt marshes, specifically for samphire cultivation (as has happened elsewhere)?

Although Britain may simply lack the coastal space for such an enterprise, it does have an abundance of farmland that is underused but owned by farmers eager to diversify their sources of income. The organic growing of elder is one response still in its early stages, while exploiting bracken rather than trying to get rid of it is another. Some landowners are finding that the economic fruits of an old meadow are the seeds of the flowers it contains. Foresters of the future may find mosses in their plantations as profitable as the trees they were trained to tend. Part of the future will certainly involve labelling schemes that certify the sustainability of products and their links with the environment. The company Direct from Dorset, set up after the Foot-and-Mouth crisis, promotes products from the county – like local charcoal – that have passed a number of criteria relating to their sourcing and processing, and bring with them environmental benefits. Kew sells its own Bar-B-Kew charcoal from the woodlands of its Loder Valley Reserve at Wakehurst Place, and the length and breadth of Britain is now rich with other so-called countryside products. Relatively minor they may all be, but these kinds of economic activities contribute to maintaining an attractive and diverse British landscape.

Charcoal from Wakehurst Place.

Schemes that label countryside products are commonplace and doubtless less scrupulous ones will try to ape them. Just how much this happens will need monitoring, as will the amounts of wild species that we use. Their sustainable use is now featuring in future plans and strategies, most recently (and potently) in the Global Strategy for Plant Conservation that was adopted by the UK in April 2002. This sets out a blueprint for how we should conserve plant life in the UK. It has five main objectives, one of which is "Using plant diversity sustainably". This use has both national and international connotations and is set out in three targets:

1 no species of wild flora endangered by international trade;

2 30 per cent of plant-based products derived from sources that are sustainably managed;

3 to halt the decline of plant resources, and associated local and indigenous livelihoods, local food security and health care.

As we go to press the 16 targets set by the Strategy are being finalised. A second edition of this book in ten years time will be able to assess how effective this will have been and tell the stories of other species whose economic value we will have suddenly come to recognise – adding yet more products to Britain's wild harvest.

Further reading

Sanderson, H. and Prendergast, H.D.V. 2002. *Commercial uses of wild and traditionally managed plants in England and Scotland*. Royal Botanic Gardens, Kew.

This book is based on research originally presented in the report above. Available on the Internet at www.kew.org/scihort/commusesreport.pdf, it contains numerous references and the contact details of many individuals and companies mentioned in *Britain's wild harvest*.

Dyke, A.J. and Newton, A.C. 1999. Commercial harvesting of wild mushrooms in Scottish forests: is it sustainable? *Scottish Forestry* 53: 77-85.

Inskipp, C. 2003. *Making a lasting impression: the impact of the UK's wildlife trade on the world's biodiversity and people*. Cambridge: WWF and TRAFFIC.

Milliken, W. and Bridgewater, S. 2001. *Flora Celtica: sustainable development of Scottish plants*. Edinburgh: Scottish Executive Central Research Unit.

Murray, M. and Simcox, H. 2003. *Use of wild living resources in the United Kingdom – a review*. A report produced for the UK Committee for IUCN.

Wong, J. and Dickinson, B. 2003. *Current status and development potential of woodland and hedgerow products in Wales*. A report commissioned by the Countryside Council for Wales, Forestry Commission and Welsh Development Agency.

Index (common and scientific names)